The Twisted Earth

The Twisted Earth

REFLECTIONS ON PATTERNS PEOPLE AND PLACES

BY HOWARD F. DE KALB

For Stu Holm

LIBRARY OF CONGRESS CATALOG CARD NUMBER: 89-81055

ISBN: 0-9623271-0-7

Printed in the United States of America
Lytel Eorthe Press, Hilo, Hawaii

First Edition

Cover by Dana De Kalb

Long ago,
    When Fu Hsi ruled the world,
He looked up,
    And considered the images of the heavens,
He looked down,
    And considered the patterns of the Earth.

Looking up,
    We examine the signs of the heavens,
Looking down,
    We examine the lines of the Earth.

        Fifth and Sixth Wing, Ta Chuan.

To heaven,
    They gave the number three,
To Earth,
    They gave the number two,
From these,
    Came the other numbers.

        Eighth Wing, Shou Kua.

CONTENTS AND ILLUSTRATIONS

CHAPTER 5 THE PLANETS

CHAPTER 6 THE APPLICATION

CHAPTER 7 THE CRITICISM

PREFACE

This book is a personal statement of ideas developed over many years and contains a number of interwoven themes. The central focus is on the observation that many structural features on the surface of the Earth appear to follow linear trends when viewed in Mercator Projection. In chapter 1 examples are given which define four principle directions that combine to form a repeating pattern of rhombus elements. The pattern also appears to be hierarchical, which means that it can be repeatedly subdivided into smaller elements that retain the same configuration. In chapter 2 a mechanical analogy is offered as an explanation in which the Earth is twisted about its axis. This shear stress generates a strain pattern in the Earth's crust similar to that observed on the surface. In the following two chapters other examples are presented that show the correspondence between certain structural features and the rhombus strain pattern. Features from the limited area of the Hawaiian Islands are first studied, after which other examples are taken from the world at large. A chapter then follows that extends the study to other objects in the Solar System, including the Moon, Mercury, and Mars, as well as several planet satellites. In the final two chapters a number of questions are raised regarding the strain analogy, and examples presented to indicate possible applications of the observation in the practical world.

A second theme woven into the discussion concerns the broad elements of geologic analysis. The ideas presented are merely one person's opinion and could differ significantly from other approaches. Emphasis is placed on the infinite complexity of the natural world in contrast to the discrete and finite process of analysis. This means that the elements used in any analysis are no more than a few items selected from the vast pool of nature. The natural world is an endless continuum that blends at the center and fades at the edges of knowledge. The subdivision of that continuum into discrete elements and categories is manmade and, necessarily, somewhat arbitrary. Selection of study items from the natural world is further constrained by the increasing use of high-speed computers which are restricted to processing huge quantities of numbers. There is a temptation to judge an analysis by the sophistication of the program and the number of calculations rather than the degree of correspondence with the real world. Mother nature is a hard taskmaster, and with each labored advance to new understandings comes a multitude of new questions. This is the frustrating fascination of nature that pervades all analysis. An awareness of the limitations and constraints surrounding the translation of nature to numbers is the foundation for all geologic analysis.

In addition to the above, infused throughout the book is a collection of odd bits and pieces of scattered knowledge related to the area under study. This includes items on weather, volcanoes, oceans, and rivers, as well as people and places. These snippets of information have no particular value as related to the central theme, but rather serve to fill in pages that otherwise might have been relatively blank. With so many happenings around the world, it seemed a pity not to make full use of the pages in the book. Hopefully, this will also make the matter a little more interesting.

The ideas presented in the central theme first took root in Iran where the oil fields follow a very obvious trend. It was there I met John Halsey, a petroleum geologist who is a true expert on fracturing. He introduced me to the concept of structural trends which eventually led to the developments in the book. The ideas were more fully developed while in Saudi Arabia, and finalized in Hilo, Hawaii after retirement. In Saudi Arabia a number of people in the Cartography Department supplied invaluable advice while I was working on the illustrations in my free time. Ibrahim Saeed Al Kalali, Ray Allen, Steve Greenough, and Randy Patrick were all both helpful and interested. It is also interesting that each of them were from a different country; Saudi Arabia, England, Australia, and the United States. My cousin Ila May Lewis from Eugene, Oregon stepped in at one stage to run errands that I could not do myself while in Saudi Arabia, and Bill Pike from the Portland Atari Club took time to change my Atari disks to IBM format. Lucretia Pladera, District Administrator of the Hawaii County library system, was kind enough to read and edit the final draft. Three others took time to read the book for content; Malcolm Smith, Director of the U.K./Canada/Netherlands Joint Astronomy Centre, Tom Wright, Director of the Hawaiian Volcano Observatory, and Charles Garcia, Superintendent of the Mauna Loa Solar Observatory, all located on Hawaii Island. Last but not least, my younger daughter Dana De Kalb took time off from her unique and beautiful sculptured painting to letter all the illustrations and paint the cover. Many thanks to all those mentioned above, and also to others who have quietly listened to me rave on about directionals. However, even with all the help there are undoubtedly many mistakes for which I must take full credit. It is also stressed that the book is a personal statement, and none of those mentioned necessarily concur with the ideas presented. English Units are used throughout simply because I've always used them and they have meaning to me. Hopefully, it will not overly distract the many who use the metric system.

1 THE DISCOVERY

FINDING THE PATTERN

Geologic analysis can be quite different from that of other sciences. In physics and chemistry experiments are made and results recorded, then tested against a theoretical model. In the biological sciences the experiments may take somewhat longer, but the results can still be seen during a lifetime. However, in geology there is little that can be done on the experimental side because of the extended time and huge forces involved. The present is all there is, and the past and future must be inferred. Rarely is there one answer, although some might try to make you believe so. Is it the slow grinding forces over millions of years that have changed the Earth, or rather, perhaps just a few major catastrophes? The answer is neither and both. There are certainly selected features that could only have been formed by some major catastrophe, but the Earth is also covered with others that can only be explained through slow erosion and depositional forces. And in the end, there will never be a final proof because the only available evidence is circumstantial.

Analysis in geology must start with a description of what exists today because that is all we truly know, and sometimes even today is questionable. How often is heard, "That shouldn't have happened," or "This is the way it should be," or "That just can't possibly be right." Each statement implying truth contradictory to fact. With the growing proliferation of complicated and sophisticated mathematical models, there is a tendency to focus on the model and exclude reality. We sometimes forget that in comparison to the continuum of the real and uncertain world, the discrete model composed of selective and incomplete interpretations is always wrong in some

respects. The question to be asked is, "How wrong and in what way, and to what limited set of observations can the model be used to approximate nature?" There is no one analysis or one answer in geology. Each method contributes to the overall understanding of the processes at work, but can never provide a complete picture of all aspects of an indivisible nature.

As explained above, geology has two facets, first is the description of the present, and second the inference to the unknown past and future. A good example of this is a comparison of Physical Geology and Plate Tectonics. Physical Geology is a study of existing land forms in an effort to provide a description of present day Earth features. On the other hand, Plate Tectonics attempts to explain the past, present, and future by means of moving portions of the Earth's surface. The analysis that follows is of the first type and deals with a description of the present. More specifically, a pattern, or framework, is presented that seems to follow the grain of the Earth's surface. No attempt will be made to carry the analysis backward or forward in time. As in much geologic analysis, the development of the pattern will take the form of examples. These will be used, first to define the pattern, and then to further illustrate how it meshes with various surface features on the Earth. It will be found that the pattern consists of straight lines that reflect in various ways the configuration of certain features.

In defining the pattern, only one type of feature will be used in order to simplify the analysis and to reduce bias. Coastlines represent one of the simplest of Earth features. They are merely a line with water on one side and land on the other, and can be drawn directly from a map without need for adjustment or interpretation. This brings up the subject of map projection. The search will be for a pattern composed of straight lines, so that a logical choice is Mercator Projection on which any straight line maintains a constant direction. Other properties such as area and distance are distorted, but this is relatively unimpor-

tant as related to the particular aspects under study. The examples used are traced from the Mercator Projection maps published by the Defense Mapping Agency of the United States Department of Defense. All coastlines drawn from these maps will show a heavy line on the water side and a finer line on the land side.

The search for the pattern will take the following form, first find a relatively long section of coastline that maintains a constant direction, then measure the direction angle of that coastline in degrees to define a "directional." Other examples will then be used to further illustrate the directional. This process will define a set of directionals that will form a pattern. The first chapter will be devoted to this empirical search for a pattern, and in the second chapter a possible explanation will be developed.

1-1 N40W Directional

This shows the coastlines used to define and illustrate the first directional. These coastlines will be discussed from left to right, generally corresponding to a west to east movement across a world map with north directed up. On the left is the west coastline of North America shown in two parts. From north to south it follows the coast of lower Alaska and Canada down to Vancouver, located just above the United States border. The section from Vancouver to San Francisco has been omitted but will be discussed later. The coastline continues south from San Francisco to the tip of Baja California in Mexico, maintaining a constant direction. East across the Gulf of California is the coastline of Sonora and Sinaloa in Mexico, oriented in the same direction which measures north 40 deg west. This method of giving a direction means that a line positioned straight up north is rotated 40 deg counterclockwise to the west. The measured direction defines the first directional that will be called the N40W directional, or sometimes just N40W so the word directional will not have to be endlessly repeated.

East across North America and the Atlantic lies Africa. Just above the equator on the western protrusion of northern Africa is an 800 mile stretch of coastline that follows the N40W directional. It runs from Dakar, Senegal south to Cape Palmas on the Liberia-Ivory Coast border. The southern part of Liberia is called the Grain Coast from the Melequete pepper seeds that were exported from there in past centuries. North of Africa in the eastern Mediterranean is Greece on the Balkan Peninsula. Two N40W directionals trace the coast on both sides of the peninsula. The area is well known for its volcanic and earthquake activity. Thera Island, also called Santorini, lies 100 miles to the southeast. One of the largest historic volcanic eruptions occurred there about 3,600 years ago. The explosion could have been heard all over Europe, and total darkness probably covered the area around the island for several days. The destructive ash falls and flooding that followed may have been a major factor in the decline of the Minoan civilization on nearby Crete.

Further east are Pakistan and India bordering the Arabian Sea near the north edge of the Torrid Zone which is marked by the Tropic of Cancer. Between the two countries is the coastal salt marsh called the Rann of Kutch, probably meaning desert of Catechu Acacia trees. Resin from the tree was used for tanning and dyeing in earlier times. The 500 miles of coast from Karachi, Pakistan south to the Gulf of Cambay follows the N40W directional. East of India and the Bay of Bengal is Southeast Asia with its many islands and peninsulas. The Malay Peninsula, once called Chersonesus Aureg, the Golden Peninsula, lies just above the equator on the Strait of Malacca. From Singapore at the southern tip, north to the Isthmus of Kra at the knee is about 500 miles. The coastlines both above and below the central offset follow N40W. South across the Strait of Malacca lies Sumatra, the westernmost island in the Indonesian island chain that stretches for 3,500 miles across Southeast Asia. The southwest coast of Sumatra follows N40W for about 1,000 miles. Parallel to the coastline and just offshore is the 25,000 foot deep Java Trench, and just inland the Sumatra-Java Mountain Range.

North of the Malay Peninsula is the Indochina Peninsula with Laos, Cambodia, and Indochina occupying the east portion, which is dominated by the Mekong River Delta on the South China Sea. This region is one of the most prolific rice growing areas in the world. From the Mekong River Delta northwest to Thailand the coastline follows N40W. East of Indochina is New Guinea, the second largest island in the world with mountains over 15,000 feet high. The eastern half of the island is the independent nation of Papua, formerly an Australian territory. The eastern part of Papua consists of a broad tail about 500 miles long trending N40W. South of New Guinea across the Coral Sea is the northeast coast of Australia that parallels the Great Barrier Reef just offshore. This is the longest coral reef in the world, stretching for more than 1,200 miles. About 500 miles of this coastline from Townsville south to Brisbane follows the N40W directional

The last two examples are far to the north in the Arctic. The northeast one-third of Russia is called Siberia. It has the greatest temperature extremes in the world, ranging from -90 to $+100$ degrees Fahrenheit. Northeast of Siberia on the Arctic Circle lies the Chukchi Peninsula, just across the Bering Strait from Alaska. The coastline of the peninsula runs for about 500 miles and trends N40W on either side of a central offset. North of Hudson Bay in Canada is Baffin Island, discovered by Martin Frobisher in 1576 when searching for the Northwest Passage around North America. This large island which lies across the Arctic Circle is about the size of Sumatra. Its rugged northeast coastline follows the N40W directional for about 800 miles.

These few examples merely indicate the worldwide distribution of the N40W trend along major coastlines. They establish the existence of what appears to be a major linear trend across the surface of the Earth that has been defined as a N40W directional. On the following pages other widely distributed linear trends will be examined

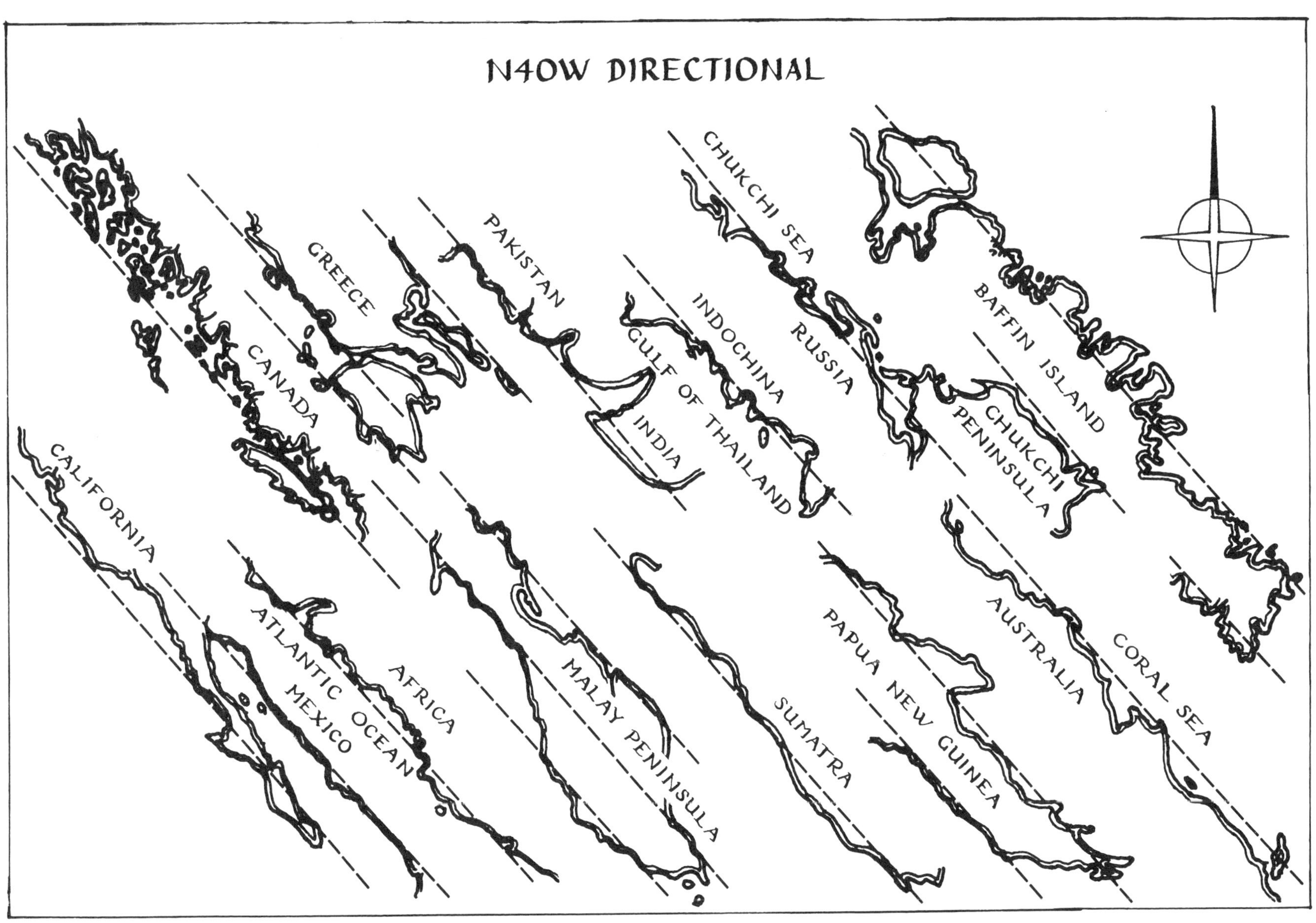

N40W DIRECTIONAL
CHUKCHI SEA
PAKISTAN
GREECE
CANADA
INDOCHINA
GULF OF THAILAND
INDIA
RUSSIA
BAFFIN ISLAND
CHUKCHI PENINSULA
CALIFORNIA
ATLANTIC OCEAN
MEXICO
AFRICA
MALAY PENINSULA
SUMATRA
PAPUA NEW GUINEA
AUSTRALIA
CORAL SEA

1-2 N50E Directional

This shows the coastline that defines the second directional. As on the previous page, the description will be generally left to right, representing a movement from west to east around the globe. On the left is the east coastline of North America, running from Nova Scotia in the north, south to Florida. The coastline is rugged with many inlets, bays, and capes, giving it the excellent harbors that have generated the highly urbanized and industrialized central area from Boston south to Washington D.C. The New York-New Jersey port alone can berth almost 400 ships at one time. This central area is inset to the northwest relative to the Nova Scotia and Carolina coastlines on the north and south that lie along the same linear trend. The direction of the trend is north 50 deg east, representing a 50 deg rotation from north to east. This defines the N50E directional that will sometimes, as before, be called simply N50E.

To the north is Alaska with the Alaskan Peninsula trailing out toward the southwest. This coastline also follows N50E, from Anchorage in the northeast along Kodiak Island to Unimak, Unalaska, and Unmak Islands beyond the peninsula to the southwest. The area is prone to severe earthquake and vocanic activity. One of the greatest earthquakes recorded, with a magnitude of 8.9, struck the region in March 1964, causing enormous damage to southern Alaska. In addition, the heaving sea floor started a huge tsunami wave that caused much destruction as it travelled across the Pacific. One of the most violent volcanic eruptions in history occurred here in 1912 when the volcano Katmai on the Alaska Peninsula erupted and filled a nearby area with ten square miles of glowing ash, creating the Valley of Ten Thousand Smokes.

East across the Atlantic is the European coastline that borders the English Channel and trends N50E along the coasts of Holland, Belgium, and northern France. Holland, on the north, has built up many miles of dikes over many years to create new polder land from the sea floor. However, this has led to severe flooding when the raging North Sea breaks through a dike. After disastrous floods in 1953, construction began on the largest sea dam in the world that now stretches for more than 20 miles across the mouth of the Zuider Zee. It is hoped that this will lessen the threat of flooding. Spain lies on the Iberian Peninsula. Its southeast coast borders the western Mediterranean, running for about 500 miles from the Costa Brava above Barcelona south to Cabo de Palos, where Columbus left Europe in 1492 to sail into the unknown and discover the Americas. Both the central offset and the north and south extremities trend along N50E. Southeast of Europe across the Mediterranean and Arabia is the horn of Africa that juts out into the Arabian Sea. Somalia, with one of the harshest climates in the world, occupies most of the area. The year round average mean temperature for much of Somalia is over 85 degrees Fahrenheit. This smooth 800 mile coastline also trends N50E. East of Africa is the triangular peninsula of India. The eastern coastline runs for more than 600 miles along N50E from Calcutta south to the Krishna River which borders the state of Orissia. Some of the worst famines in history occurred here in the late 1870's when the monsoons failed and millions were lost to starvation.

Northeast of India is the huge country of China that occupies much of eastern Asia. On the east coast of China is the Gulf of Chihli, also called Bohai, that serves as a sea outlet for nearby Beijing, the capital of China. Although the land is fertile and good for crops, it has seen some of the worst floods and earthquakes in history. The Tangshan earthquake in 1976 may have caused three-quarters of a million deaths. The predominate coastline trend here is again N50E. Further north up the coast of Asia, is the Russian Okhotsk Sea and the Gulf of Selikova. Both the coastline bordering the sea in the south and the gulf in the north follow N50E. The island of Borneo, third largest in the world, lies on the equator to the south.

The northwest coast of Borneo runs along Sabah on the north and Sarawak on the south, which are both part of Malaysia. Between them is oil-rich Brunei, an independent sultanate. The coastline follows N50E for about 600 miles along the South China Sea. South of Borneo and the equator is the continent of Australia. The Kimberley coast on the northwest is dry and desolate, with a highly variable rainfall season of only three months. This coastline runs for about 400 miles following N50E.

Southeast of Australia is New Zealand, that the Polynesians say was pulled up from the ocean by the great hero Maui using the sky fishhook we know as Scorpio. The northwest coast of South Island is about 500 miles long, and parallels the spine of the Southern Alps and the Alpine Fault. It also trends N50E. Far to the south is the Antarctic Peninsula on the Antarctic Circle, with more than 20 research stations maintained by Argentina, Chile, and the United Kingdom. It resembles an inversion of the Alaskan Peninsula, again trending N50E. It is interesting to note that the two defined directionals of N40W and N50E intersect at 90 deg, a right angle. This is the first indication of a certain regularity in the linears on the Earth's surface, indicating the possibility of a pattern. On the next pages, two more directionals will be defined and illustrated to complete the pattern orientations.

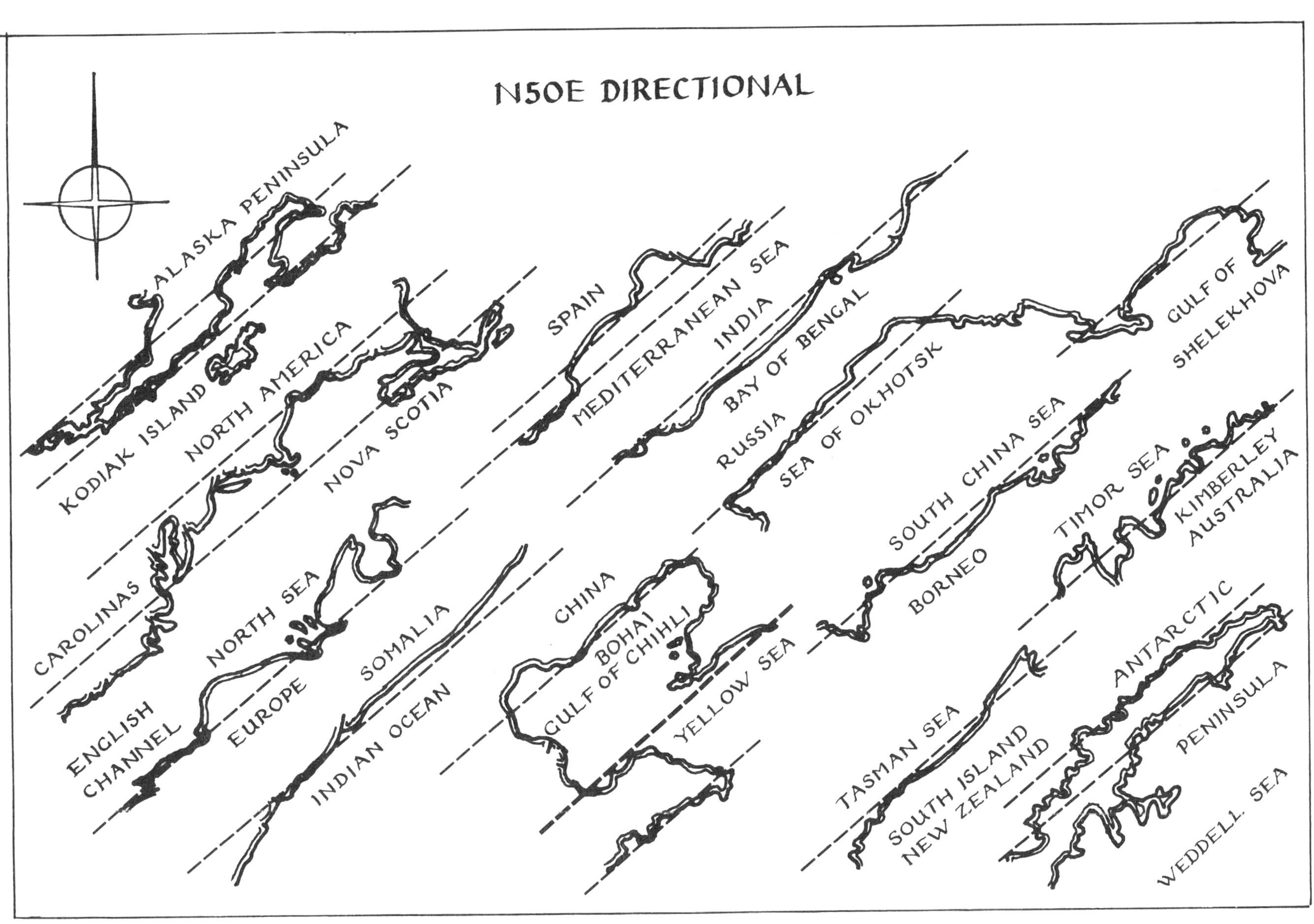

N50E DIRECTIONAL
ALASKA PENINSULA
KODIAK ISLAND
NORTH AMERICA
NOVA SCOTIA
CAROLINAS
NORTH SEA
ENGLISH CHANNEL
EUROPE
SOMALIA
INDIAN OCEAN
SPAIN
MEDITERRANEAN SEA
INDIA
BAY OF BENGAL
RUSSIA
SEA OF OKHOTSK
SOUTH CHINA SEA
BORNEO
TIMOR SEA
GULF OF SHELEKHOVA
KIMBERLEY
AUSTRALIA
CHINA
BOHAI
GULF OF CHIHLI
YELLOW SEA
TASMAN SEA
SOUTH ISLAND
NEW ZEALAND
ANTARCTIC
PENINSULA
WEDDELL SEA

1-3 N7E Directional

This shows the coastline examples that follow the third directional. On the left is the west coast of South America which stretches for 3,000 miles along the Chilean coast from Peru south to Tierra del Fuego. The Peru-Chile Trench, which reaches a depth of 26,000 feet, lies just offshore. Inland are the Andes with Liullaillaco rising to 22,000 feet, the highest volcano in the world. To the north is the Desert of Atacama, the driest place in the world where in some places rainfall has never been recorded. The coast forms part of the volcano and earthquake Ring of Fire which circles the Pacific Ocean. The May 1960 Conception earthquake in Chile may be the greatest ever recorded. Sea floor movement produced a giant tsunami wave that moved across the Pacific and carried Japanese fishing boats 150 feet inland. This linear stretch of the Chilean coast has a measured direction of north 7 deg east which defines the third N7E directional.

North of South America is the west coast of North America which trends N40W, except for an offset along the coasts of Oregon and Washington. Here the coastline follows the N7E directional from Vancouver Island in Canada south to Northern California, a distance of about 500 miles. Behind this coast is the Cascade Range which is also part of the Ring of Fire. In May 1980 Mount St. Helens erupted and lost 1,200 feet of height as it blasted away 150 square miles of forest. East of the Cascades in Washington State are the Channeled Scablands, formed when the Lake Missoula glacial dam burst several thousand years ago at the end of the last ice age. A catastrophic flood up to 800 feet deep may have run for two weeks. It cut down 400 feet into solid basalt forming the coulees, or dry valleys, and stripped off topsoil over a 2,000 square mile area.

Between North and South America is the Yucatan Peninsula in Mexico that juts out into the Gulf of Mexico. When Cortes crossed the area in 1525 he found the ruins of the huge Mayan temple complexes of Mayapan, Chichen Itza, and Uxmal. This coastline generally trends N7E with an offset near the center. North of Yucatan is the western Gulf Coast which runs from Corpus Christi, Texas south to Tampico, Mexico, following the N7E directional. The area is often hit by hurricanes which are considered to be the greatest storms on earth. The Great Hurricane of 1874 destroyed all shipping on the Rio Grande, and more recently in September 1967, Hurricane Beulah with its tornadoes and floods caused a billion dollars damage in The United States and Mexico.

Far to the northeast above the Arctic Circle is Spitzbergen, or officially Svalbard, meaning the cold coast. It was discovered by Icelanders in the 12th century, then was forgotten for 400 years before its rediscovery. Both Norway and Russia have permanent settlements there to work the coal mines. The southeast coast of Spitsbergen trends N7E with an offset at the center. To the south is the Scandanavian Peninsula, with Sweden on the east bordering the Baltic Sea and Gulf of Bothnia. At the northern reach of the Gulf of Bothnia is Torino, Finland, just south of the Arctic Circle. In 1736 the Frenchman Maupertuis led an expedition there to measure a degree of latitude. The results proved that the Earth bulged at the equator as Issac Newton had predicted. The 400 mile Swedish coastline follows the N7E directional. The west coastline of the Iberian Peninsula on the Atlantic runs for 400 miles along the N7E directional. At the extreme south of the peninsula is the Portuguese Cape of Sao Vincente where Prince Henry the Navigator came in 1419 and sent his ships down the unknown coast of Africa. He was responsible for the discovery of the Cape of Good Hope which opened the sea passage to India.

East of the Asian continent above Japan lies the 500 mile long Island of Sakhalin. It has been a Russian possession since 1945, although Japan controlled it from time to time in the past. The island trends N7E, but has a central offset as in other examples. South of Sakhalin are the Japanese Islands which lie on the Ring of Fire. The central and largest of the islands is Honshu with a Pacific coastline trending N7E. Tokyo in the Kwanto Plain of southern Honshu was hit in 1923 with the most destructive earthquake ever recorded. The damage was estimated at three billion dollars and 140,000 people lost their lives. South of Japan in southeast Asia is the Malay Peninsula which trends N7E along its upper part above the knee. This narrow neck is called the Isthmus of Kra and is only 40 miles wide at its narrowest point. The possibility of building a canal through the Isthmus has been discussed for many years, but no firm plans have ever been made.

Southeast of the Malay Peninsula is the large island of Borneo. The east coastline generally trends N7E, if the two large bulges are ignored. These bulges will be considered later when the shape of the entire island will be examined. Southeast of Borneo below the equator is Australia with the large Gulf of Carpenteria cut out of its north coast. East of the Gulf is the Cape York Peninsula coastline which runs for 400 miles along N7E. About 20 percent of the world's bauxite production is mined at Weipa on the peninsula. On the following page the fourth and final directional will be defined and illustrated. Examples will then be presented that include combinations of all four directionals.

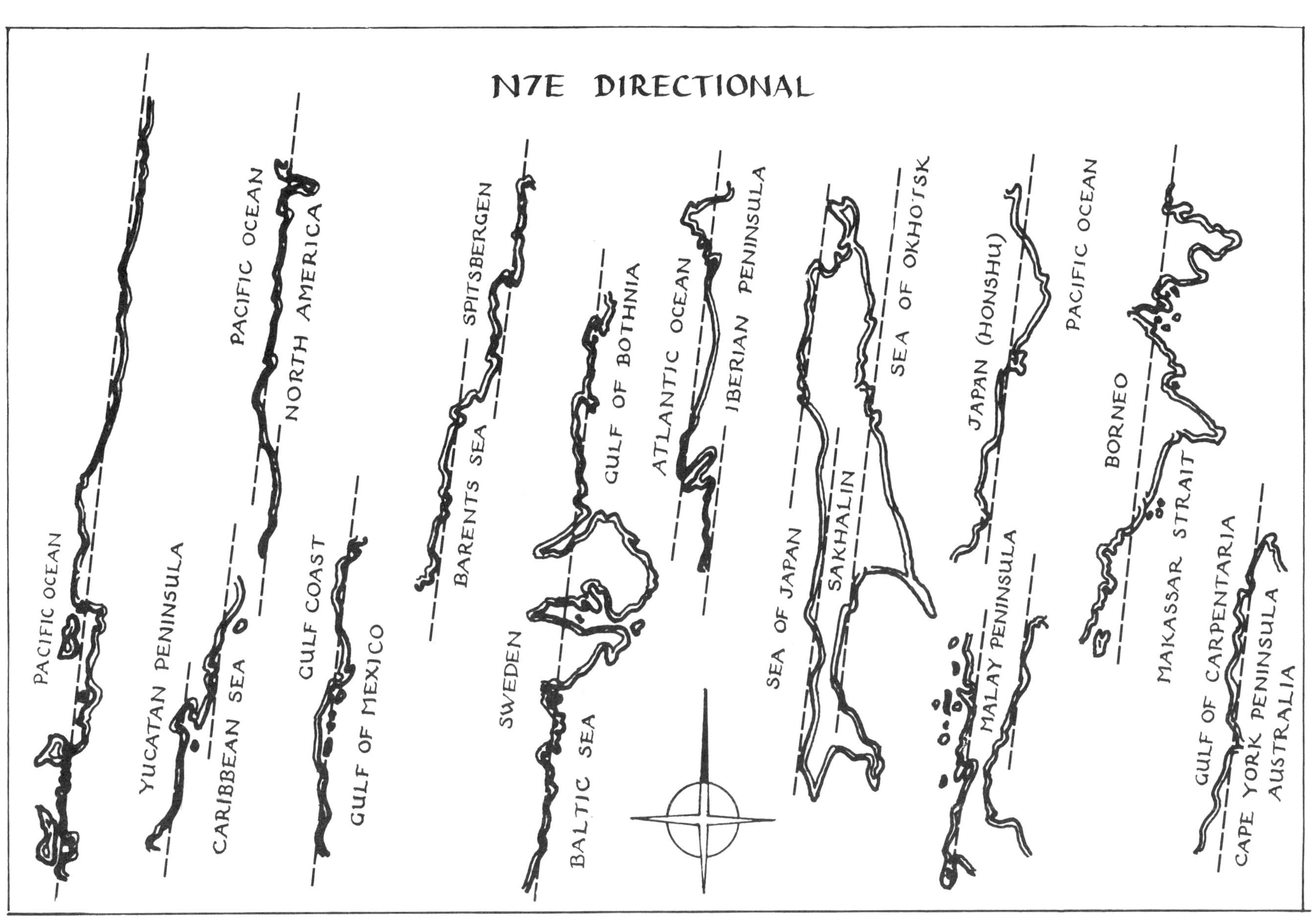

N7E DIRECTIONAL
PACIFIC OCEAN
PACIFIC OCEAN
NORTH AMERICA
YUCATAN PENINSULA
CARIBBEAN SEA
GULF COAST
GULF OF MEXICO
SPITSBERGEN
BARENTS SEA
GULF OF BOTHNIA
SWEDEN
BALTIC SEA
ATLANTIC OCEAN
IBERIAN PENINSULA
SEA OF JAPAN
SAKHALIN
SEA OF OKHOTSK
JAPAN (HONSHU)
MALAY PENINSULA
PACIFIC OCEAN
BORNEO
MAKASSAR STRAIT
GULF OF CARPENTARIA
CAPE YORK PENINSULA
AUSTRALIA

1-4 W2N Directional

This shows the examples used for the fourth and last directional. North of the American Continents just a few degrees south of the North Pole in the Canadian Arctic lie the Queen Elizabeth Islands . A strip of ocean runs east-west below these islands and separates them from another group of islands to the south. The western part of this strip of ocean is the M'clure Strait which lies below Melville Island. The southern coastline of this island is oriented approximately west 2 deg north, which means a 2 deg clockwise rotation of a line pointed due west. This defines a fourth directional which again will often be called simply W2N. The ice chocked M'clure Strait was the bottleneck that stopped all 19th Century efforts to find a Northwest Passage above Canada. Norway's Roald Amundsen finally made the voyage in 1906 by sailing south and by-passing the strait.

South of the Canadian Arctic Islands is the barren eastern half of Canada's north coast. A tongue of cold air protrudes into this area in the winter making it one of the coldest and least populated places in continental Canada. This sparsely settled arctic coastline trends along W2N. Looking south across Canada and the United States is the northern coastline of the Gulf of Mexico that stretches from Texas on the west, to Florida on the east. At the center there is an offset in the coastline at the Mississippi delta where the river enters the gulf. Both east and west of the delta offset the coastline generally follows the W2N directional. Hurricane Camille, the greatest storm North American has ever seen, smashed into the delta region in August 1969. Winds raged at 200 miles an hour, and the surge drove water to heights of 20 feet. Damage from the storm was estimated at 1.5 billion dollars.

Southeast of the Gulf Coast is the island of Hispaniola, one of the Greater Antilles, which borders the Caribbean Sea. The island is divided in two parts with Haiti on the west and the Dominican Republic on the east. Columbus discovered the island in 1492 on his first voyage across the Atlantic, and mapped the W2N trending southwest coastline in 1494 on his second voyage. Venezuela in South America, which Columbus found on his third voyage in 1495, borders the Caribbean on the south. With the discovery of oil in 1913, Venezuela developed into one of the world's major petroleum producers. The Carribean coastline of Venezuela can be extended east along the coast of Trinidad, generally following W2N. Further south on the southern tip of South America is Tierra del Fuego and Cape Horn. This cold, windy, and foggy region is the southernmost populated place in the world with a coastline that trends W2N.

On the right side of the figure at the top is the north coastline of the Iberian Peninsula oriented along the W2N directional. The Galicians, who speak a Portuguese dialect, live on the west, and to the east live the Basques, who speak a language unrelated to any other known tongue. Southeast across Europe and the Middle East is the north coastline of the Arabian Sea which runs W2N along the south coasts of Iran and Pakistan. This 600 mile coast is dry, barren, and desolate, but also with widespread flooding when the infrequent rains do appear. Alexander's army marched along the coast in 325 B.C. on their way home to Greece from India. Far to the northeast across Asia is the arctic coast of Russia, lying just west of the Kamchatka Peninsula and bordering the Okhotsk Sea. The land along this W2N trending coastline is high and drops away toward the interior, so that rivers flow north away from the sea, building up a thick arctic tundra that can range from 1,500 to 4,500 feet in thickness.

South of the Okhotsk Sea is Japan, with the largest island Honshu located in the center of the island chain. On the previous page the east coast of Honshu was found to follow the N7E directional, but further south the island bends westward to follow the W2N directional. As previously noted, Japan lies on the Ring of Fire and has had many severe earthquakes. In 1855 a major earthquake hit Tokyo, then called Edo, and virtually destroyed the city. Coincidentally, the quake occurred just after Commodore Perry's visits of 1853 and 1854 which opened Japan to the outside world. Quite naturally, there were many who linked the two and blamed the earthquake on Perry's intrusion. South of Japan on the equator are the Celebes in Indonesia. The island is made up of a central body and four arms, forming what might be called a quadrapus rather than an octopus. The upper arm, called the Minahassa Peninsula, extends north from the central body and then abruptly turns east to follow N2N. Although Indonesia is a Moslem nation, most of the Celebes people are Christians. Below the Celebes is Java, also in Indonesia. It has very fertile volcanic soil and is one of the most densely populated places in the world. The island generally trends W2N with the center offset.

It will be noticed that the W2N examples are less clear-cut than the first three directionals. In order of clarity, the first two, N40W and N50E, are probably most clear, followed by N7E, and last, W2N. This observation will be discussed in Chapter Two. The next two figures will show coastline examples that include all four directionals.

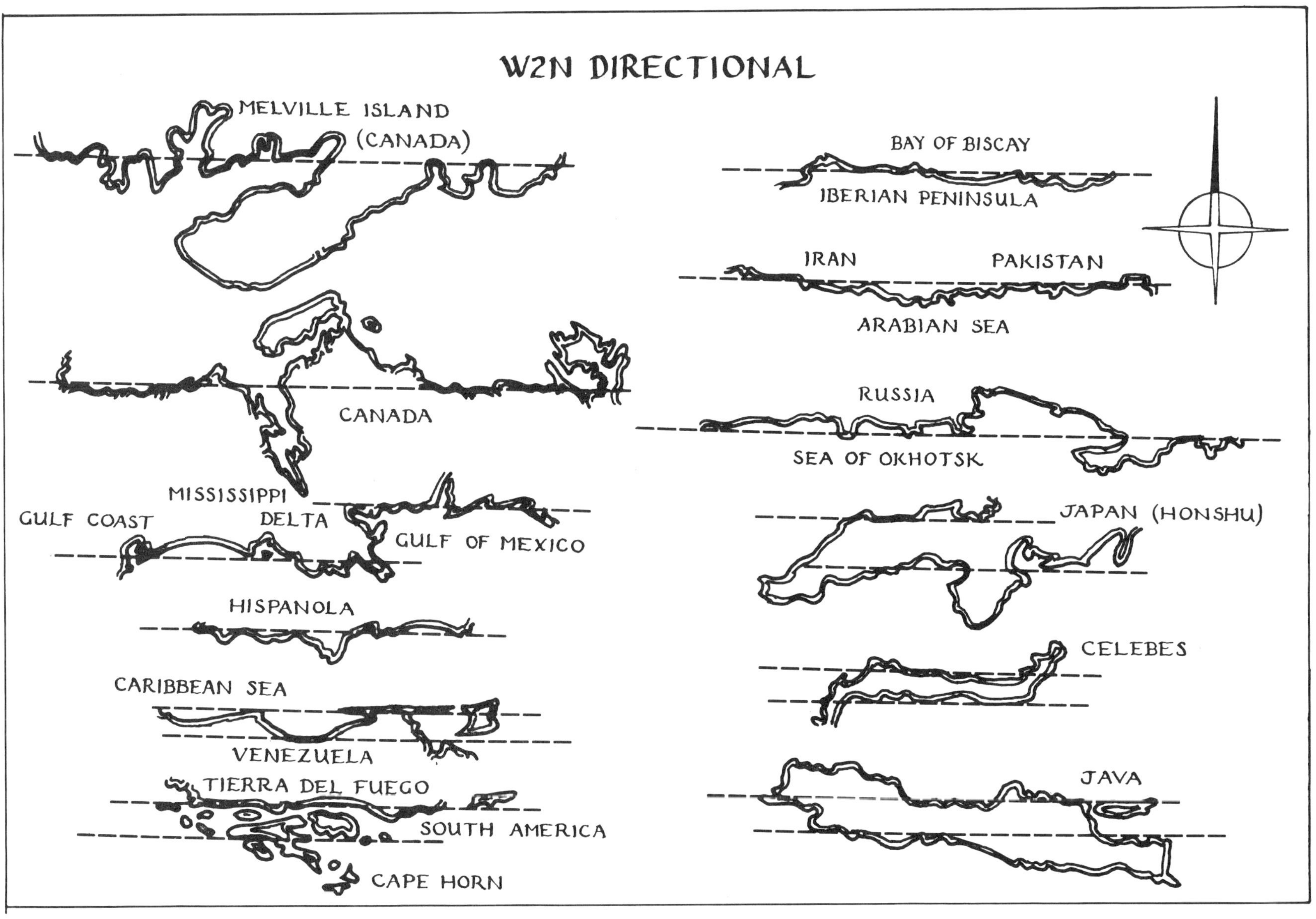
W2N DIRECTIONAL
MELVILLE ISLAND
(CANADA)
BAY OF BISCAY
IBERIAN PENINSULA
IRAN
PAKISTAN
ARABIAN SEA
CANADA
RUSSIA
SEA OF OKHOTSK
MISSISSIPPI
DELTA
GULF COAST
GULF OF MEXICO
JAPAN (HONSHU)
HISPANOLA
CARIBBEAN SEA
CELEBES
VENEZUELA
TIERRA DEL FUEGO
JAVA
SOUTH AMERICA
CAPE HORN

1-5 Directional Combinations I

This shows several coastlines around the world that can be traced using a combination of all four of the directionals that were illustrated in the preceding pages. These four directionals represent four different orientations of straight lines that tend to follow coastlines on a Mercator projection of the Earth. The four directionals are N40W and N50E that intersect in a right angle and are most clearly defined, N7E that is somewhat less distinct than the first two, and W2N that is least clearly defined of the four directionals. This observed degree of clarity will also be found in the examples on the following pages. As before, the coastlines will be described from left to right. The examples include both previously given and new coastlines.

In the far arctic on the north edge of the Eurasian Continent is the Taymyr Peninsula, flanked on the northwest and southeast by the Kara and Laptev Seas. It is a cold, barren wasteland of tundra with only a very small population of nomadic reindeer herders. The Kara Sea to the northwest is icebound, and effectively stopped the many attempts, beginning in the 16th century, to find a Northeast Passage around Russia. At that time Spain and Portugal controlled the sea lanes around Africa and South America. It was not until the end of the 19th century that A.J. Nordenskiold from Sweden made his way across the Arctic Seas and successfully navigated the Northeast Passage. The Taymyr Peninsula can be outlined with the first two directionals. The Kara and Laptev Sea coastlines follow the N50E directional on the northwest and southeast, and the coastline on the northeast trends along N40W, broken by a central offset.

Halfway around the world is the east coastline of the North American Continent that runs from Nova Scotia in the north to the Carolinas in the south. This coastline is bounded by two parallel N50E directionals. The first traces the Nova Scotia coast with a continuation to the southwest along the Carolinas.

Between Nova Scotia and the Carolinas, along the inset to the northwest, the coast follows the second N50E directional. The bottom of the inset is bounded by a N7E directional along Cape Charles and Delaware Bay, and the top by a W2N directional along the Bay of Fundy. At the center of the inset is Cape Cod, outlined with a W2N on the south and a N7E on the east. This coastline is a good illustration of the general pattern of many coastlines around the world that generally trend along one directional, but have insets and offsets that follow other directionals. The northeast coast of Asia is shown just below the North American coastline example. This Russian coastline, that includes two examples used previously, generally trends N50E consistent with the entire length of the east coast of Asia. The general N50E trend can be seen along the west shoreline of the Okhotsk Sea to the south and the Gulf of Selikhova to the north. In the center is a W2N offset that separates the two bodies of water. At the bottom of the figure is the South Asia coastline of Iran, Pakistan, and India that follows W2N on the left, with an abrupt change to N40W on the right. Moving to the top of the figure, the coastline of France is shown along the Bay of Biscay, trending N7E in the south and W40N in the north. This is the first coastline in the world to be accurately surveyed. Jean Cassini began the work in the late 17th century, and it was not until a century later that his great-grandson completed the mapping of all of France. The Islands of New Zealand in the South Pacific trend N50E, with a sharp nose in the north that juts off at a right angle along N40W. Other directionals show up in various offsets along the coast. In the Middle East, the Persian Gulf shows a N40W general trend, but with an abrupt shift in the south, first to W2N, then to N50E. The economy of the area is dominated by oil, with more than half of the known world reserves in the near vicinity of the gulf. The Indochina Peninsula in Southeast Asia has the same rectangular shape as the Russian Taymyr Peninsula far up in the arctic north. The coastlines of the Gulf of Tonkin on

the northeast and Gulf of Siam on the southwest follow N40W, and on the southeast the South China Sea coastline follows N50E.

The west coast of Alaska is shown on the far right of the figure. The coastline pattern is somewhat different from that usually seen because it is Mercator Projection. It will be remembered that distance and area are increasingly distorted as the poles are approached, and Alaska is in the arctic just below the North Pole. The distortion has considerably enlarged the upper triangular area above the arrowlike Seward Peninsula, making it similar to the triangular area below. The upper and lower triangles and the arrow point of Seward Peninsula are formed with N40W and N50E directionals, with the arrow shaft between two W2N directionals. Additional examples using all four directionals will be given on the following page.

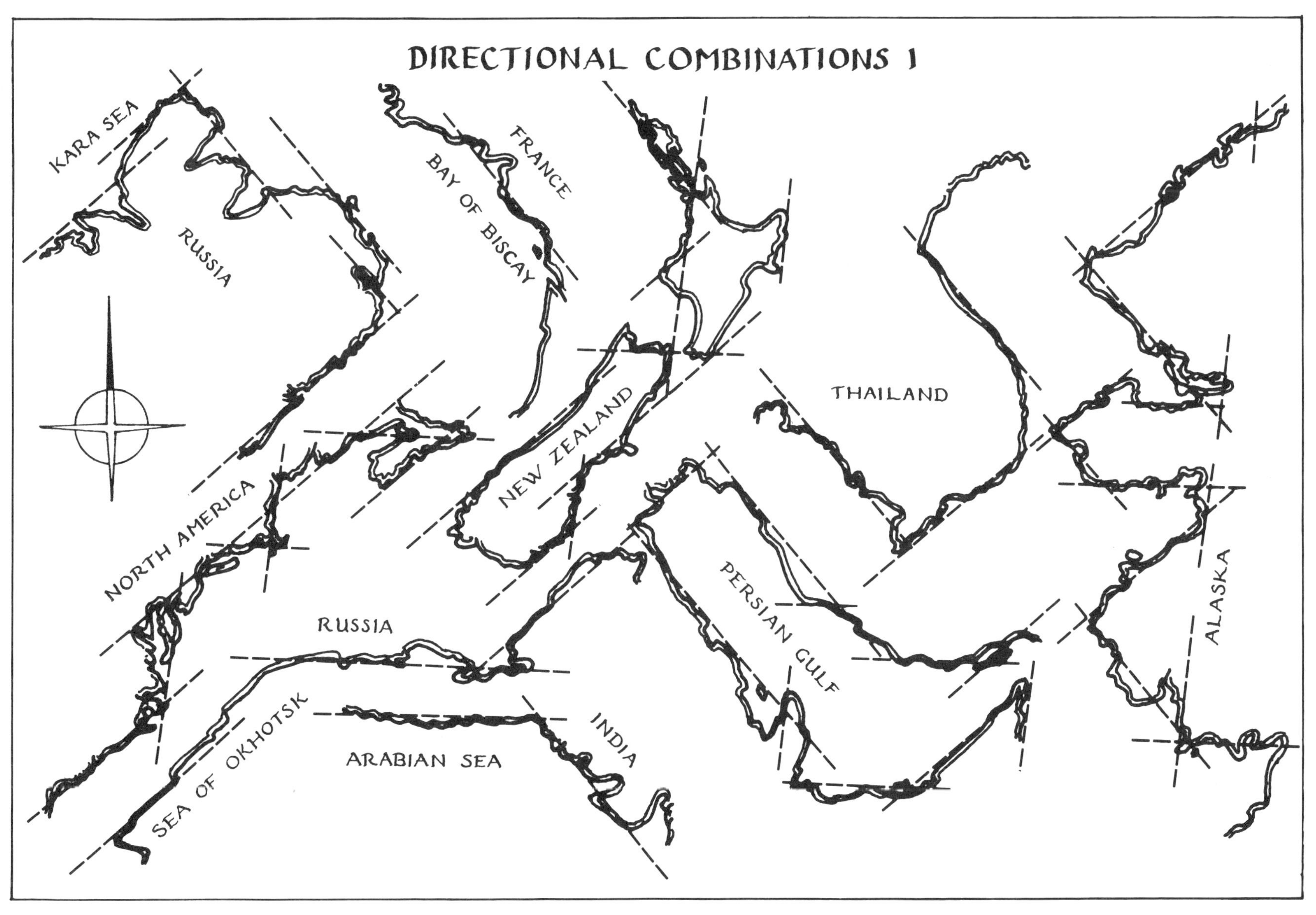

DIRECTIONAL COMBINATIONS 1
KARA SEA
RUSSIA
FRANCE
BAY OF BISCAY
THAILAND
NEW ZEALAND
NORTH AMERICA
RUSSIA
PERSIAN GULF
ALASKA
SEA OF OKHOTSK
ARABIAN SEA
INDIA

1-6 Directional Combinations II

On the left is the southern neck of South America with the Chilean coast on the west and the island of Tierra del Fuego at the southern tip. The Chilean coast strikes N7E towards the south, then shifts first to N40W, and finally to W2N along the south coast of Tierra del Fuego. Between Tierra del Fuego and the mainland is the winding Strait of Magellan that makes its way through the many islands that lie above Cape Horn. It was named for Magellan, one of the greatest seaman the world has ever known, who first made the trip through the island maze in 1520. Earlier, Magellan had sailed around Africa's Cape of Good Hope and on to southeast Asia, then returned to Spain and set out to reach the same area by sailing in the opposite direction around the Americas. He crossed the Pacific and landed in the Philippines to find people who spoke the same language as his Malay servant. Although Magellan died on the way, his deputy Sebastian completed the trip around the world and reached Spain with one ship left out of five, and only a few of the 280 seamen who started.

Next right is the Malay Peninsula, parts of which have been used in previous examples. Above the knee the Isthmus of Kra trends N7E, and below the knee the trend is N40W down to the island of Singapore at the tip. Next is shown the west coast of North America, which has been used in other examples. The Canadian coastline in the north and the coastline of Upper and Lower California in the south trend N40W, but are separated by an offset along the Oregon and Washington coasts that follows N7E. On the other side of the world in the Middle East is the 1,000 mile-long Red Sea that separates the African Continent from the Arabian Peninsula. It is fairly shallow with many submerged reefs, and has highly variable winds, making it dangerous for sailing ships. It was common practice in the ancient world to off-load ships at the south port of Aden and send the cargo overland to the Mediterranean. The Red Sea finally became an important waterway when the

Suez Canal was built by France in the mid 19th Century. Although the coastline does not match the directionals well, the contours at depth show a good fit. The dotted 3,000 foot contour line makes a stair step pattern alternating between N40W and N7E.

At the top of the page is the Iberian Peninsula which contains Portugal and Spain. It is bordered on the north, south, and west by W2N and N7E directionals making a rhombus with a missing right side. In the east on the Mediterranean the coastline follows N50E, broken by an offset at about the center. The Iberian Peninsula hangs off the southwest corner of Europe, connected only by a thin neck of land holding the Pyrenees Mountains. The Peninsula juts out south towards Africa and west into the Atlantic, and its history reflects its geographical location. The Carthaginian Hamilcar from northern Africa established the first true state in southern Spain, and 1,000 years later the North African Arabs brought science, mathematics, and craftsmanship to Spain when they invaded the country. Later, both Spain and Portugal were at the forefront in exploring Africa and the Americas.

The four main islands of Japan from north to south are Hokkaido, Honshu, Shikoku, and Kyushu. As noted earlier, Japan lies on the Ring of Fire and has experienced many severe earthquakes. Japanese legends tell of a giant catfish that lives under the Earth and wiggles its tail from time to time making the ground shake. There is also a folk belief that just before an earthquake, catfish become excited and swim to the surface from their normal resting place on the bottom. As with a number of old beliefs, it has been found to contain a measure of truth. Recent research shows that catfish seem to be highly sensitive to the many small vibrations that usually occur before a major earthquake and do become unusually excited. Along northern Honshu and Hokkaido the general trend is N7E, while south Honshu strikes off towards the west along W2N. On the extreme south, the southern tip of Honshu with Shikoku and Kyushu trends along N50E. The east coast of India is

shown from Calcutta in the north to the Coromandel Coast in the south. The trend is N50E on the north and south, with a central offset following N7E. The island of Borneo has a coastline which can be outlined with a limited number of directionals. The northwest coast trends N50E and the south coast W2N. On the east the general trend is N7E, but with two triangles of sides N40W and N50E jutting off towards the east.

The last example is the east coast of Sweden which borders the Baltic Sea and Gulf of Bothnia. As shown before, it trends N7E with the Uppsala Peninsula breaking out towards the east about a third of the way up. The peninsula coastline is triangular with the sides following N40W and N50E directionals. The Swedish Vikings left this area in the first millennium when Norsemen were raiding the British Isles and western Europe. The Swedes followed the rivers south through Russia to the Caspian and Black Seas, setting up trading posts which eventually grew up into some of the major cities of Russia. This completes the examples of directional combinations. In the next few pages the directionals will be tied together to form a pattern.

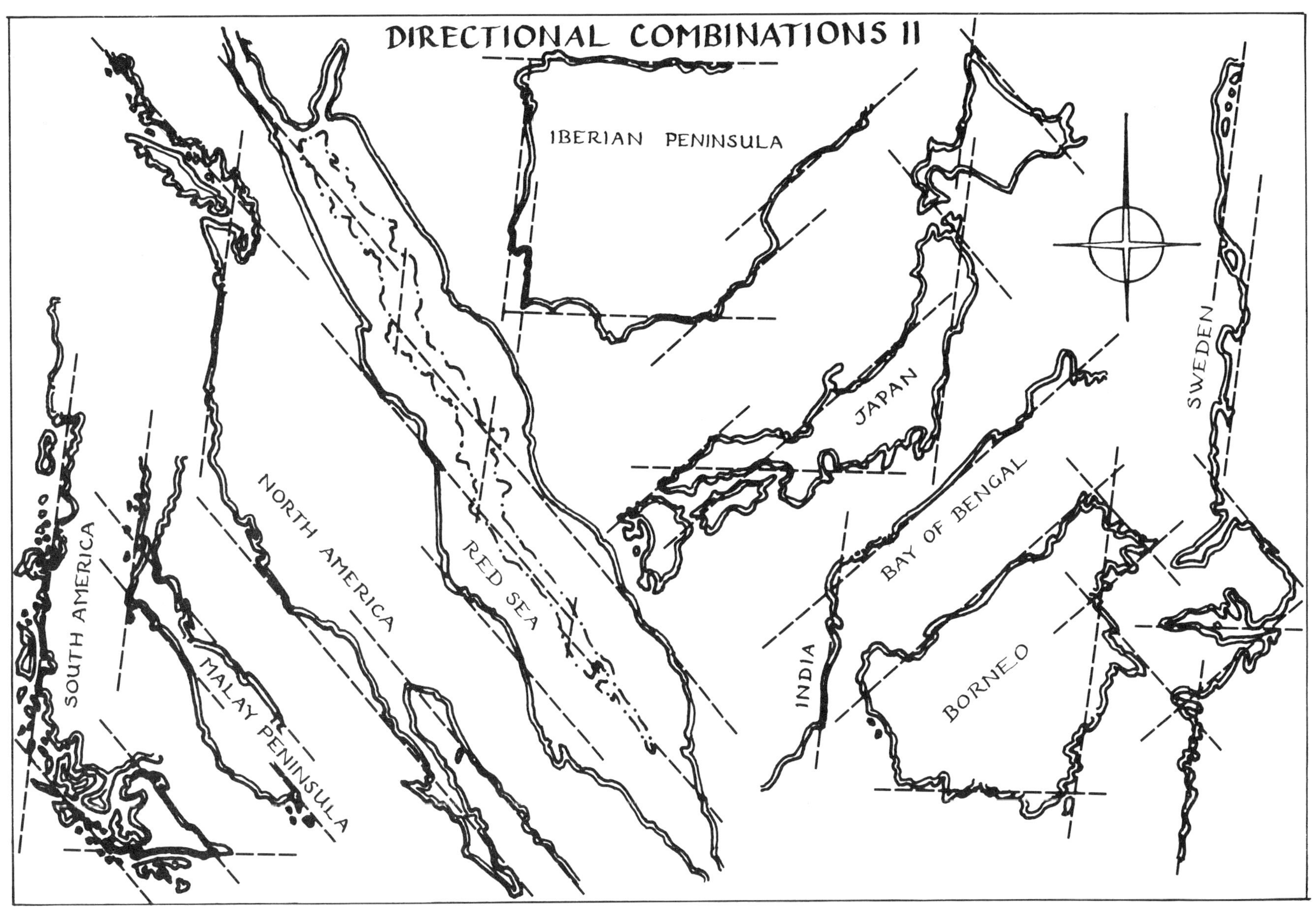
DIRECTIONAL COMBINATIONS II
IBERIAN PENINSULA
SWEDEN
JAPAN
BAY OF BENGAL
NORTH AMERICA
SOUTH AMERICA
MALAY PENINSULA
RED SEA
INDIA
BORNEO

1-7 Directional Intervals I

A geometric pattern is composed of lines arranged in some orderly way. Four sets of straight lines that follow coastlines have been defined by direction. A pattern composed of these four sets must include some sort of relationship between the different directionals. In other words, the relative position of the sets must be defined as well as direction. The relationship could take the form of spacing between the directionals, both within a set and between different sets of lines. Normally, a space, or interval, is a measure of the distance between lines. However, distance in Mercator Projection is distorted, and is more and more exaggerated as the poles are approached. The distortion is easily seen by comparing the size of Greenland in the Arctic with South America on the equator. On Mercator Projection they look the same, but in contrast, South America is actually about nine times the size of Greenland.

Lines of longitude circle the Earth, crossing the equator and coming together at the poles. Starting from the equator and moving north, any two lines of longitude draw closer together as the North Pole is approached, eventually meeting at the pole itself. In other words, the distance between lines of longitude becomes less as the pole is approached, finally decreasing to zero where they meet at the pole. However, on Mercator Projection the lines of longitude are represented by straight vertical lines which always stay the same distance apart no matter how close to the pole. If two east-west lines of equal length on Mercator Projection are compared, a line at the equator would represent a true distance about six times the distance of a line along the top of Greenland. Although distance measurement is distorted on Mercator Projection, horizontal measurement in degrees is correct. This is because a horizontal line across a Mercator world map always represents a full 360 deg circle stretched out across the map. Two horizontal lines of equal length located anywhere on a Mercator map are always equal in degrees. From now on, all map linear measurements will be given in degrees measured along a horizontal. This will allow meaningful comparisons to be made between measurements.

The north central portion of the Mediterranean Sea is shown on the figure, extending from eastern France on the left to Turkey on the right. Along this coastline are two major peninsulas and the dividing seas. The Tyrrhenian Sea on the left lies between the islands of Corsica and Sardinia to the west, and the Peninsula of Italy to the east. Further east is the Balkan Peninsula which contains Greece, and is separated from Italy by the Adriatic and Ionian Seas. East of Greece is the Aegean Sea separating Greece and Turkey, which the ancient world knew as Anatolia. At the beginning of the first millennium B.C., the Greek city-states began an expansion which took them first east to the Anatolian coast, then west to Sicily and the toe of Italy. The Greek city of Syracuse was founded in Sicily in 800 B.C. and grew to become one of the greatest cities in the ancient world. At about the same time, the seafaring Phonicians established Carthage to the southwest on the north coast of Africa, and it also developed into another great ancient city. For the next few hundred years the Greeks generally controlled the area, in spite of constant bickering between themselves, fending off attacks by Carthage on the south and the Etruscans who swarmed down from time to time from northern Italy. About the only time the Greeks acted together was in the fifth century B.C. when the Persian Darius the Great attempted the invasion of Greece from the east. He was soundly defeated on the Plains of Marathon by a combined Greek force, as was the Persian Xerxes a few years later during the great sea battle of Salamis. Perhaps it was the pride of success, along with the cultural shock and disruption, that initiated the creativity and innovation of the Greek Classical Age that occurred over the next 50 years. Of course, Imperial Rome soon entered the picture to begin the expansion that eventually covered all the Mediterranean and much of Europe. The great city of Carthage was literally wiped off the map and plowed under, and many of the Greek city-states fared little better. As the year one grew near, a series of rebellions swept through Greece, and the Roman Legions swarmed across the land putting down disorders. Both overlord and rebel seemed to cooperate in the destruction that followed, and the ruins seen today of that great Classical Greek period look about the same as they did 2,000 years ago. Perhaps creative man must eventually be balanced by destructive man.

However, nature rarely changes the shape of the land that quickly, and the coast of southern Europe probably looks much the same as it did several thousand years ago. Along the coastline the general trend of N40W is shown by dashed lines which follow the borders of the peninsulas and separating seas. It is observed that these N40W directionals seem to be equally spaced. It will be remembered that measurement is by degrees along a horizontal on Mercator Projection, so that the directionals are separated by equal intervals of 2.8 deg. Of course, the equal intervals might also be considered 5.6 deg, which is twice the smaller interval. On the following pages other examples will be studied to determine if the 2.8 or 5.6 deg intervals turn up in other parts of the world.

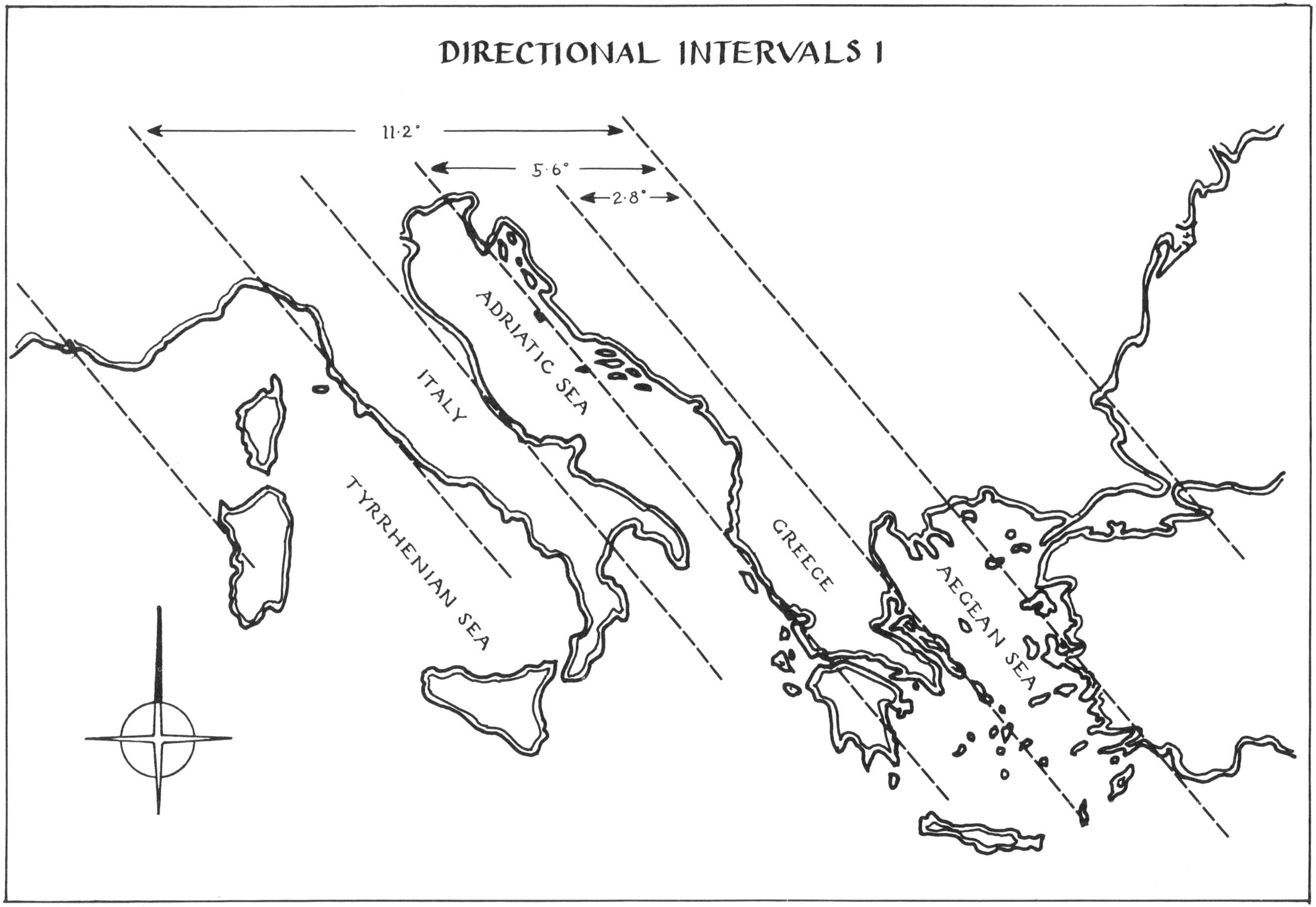

DIRECTIONAL INTERVALS I
11·2°
5·6°
2·8°
ITALY
ADRIATIC SEA
TYRRHENIAN SEA
GREECE
AEGEAN SEA

1-8 Directional Intervals II

Three examples are shown reflecting the 2.8 deg interval, or its double 5.6 deg, that was observed in the Mediterranean. On the left is the west coast of North America from Canada in the north to Mexico in the south that was previously discussed as a directional combination. It will be remembered that the general trend is N40W, as reflected in two N40W directionals that follow the Canadian coast in the north and the California coast in the south. These two directionals are separated by an N7E offset in the coastline along the states of Washington and Oregon. The offset interval measures 11.2 deg along the horizontal, which is equivalent to 5.6 deg doubled, or 2.8 deg doubled twice.

This coastline was not accurately mapped until the end of the 18th century. Although Spain had controlled California for some time, it was considered unimportant until the Russians began approaching from the north which forced Spain to colonize the country. Twenty-one missions were built along the coast and settlements grew up around them. Further north, the Russian presence also pushed England into activity, and Captain James Cook was sent into the Pacific to map coastlines and look for the Northwest Passage around the top of North America. Captain Cook was the first to develop accurate techniques of coastline surveying from the ocean. It has been said that all islands discovered by Cook were accurately placed on a map and never had to be rediscovered. He was the first to use triangulation when mapping a coastline, combining both land and sea surveying methods. Seamen were sent ashore, even on the roughest coast, to locate three accurate positions by survey. He then sailed along the coast, drawing in the coastline between the surveyed locations. He made three voyages into the Pacific, covering the ocean from the antarctic to the arctic. He left England on his third trip in 1776 to discover the Hawaiian Islands, and then to map 3,000 miles of the North American west coast from Oregon north to Alaska. Sailing with Cook was George Vancouver who continued the work after Cook was killed in Hawaii. Over a three year period at the end of the 18th century, Vancouver developed a complete map of the Hawaiian Islands, and the first accurate chart of the Pacific Coast from Lower California north to Cook Inlet, Alaska.

The example in the center of the figure is Baffin Island. It was named for William Baffin who discovered it while looking for the Northwest Passage during the early 17th century. He explored Baffin Bay and the south coast of Baffin Island, and then sailed further north to the Queen Elizabeth Islands where he found the channels filled with pack ice. Baffin was not believed when he returned with stories of his explorations, and it was some 200 years later before a ship returned to rediscover his arctic islands. Baffin died in the Strait of Hormuz at the mouth of the hot Persian Gulf, a half a world away from the land of his frozen discoveries. The general trend of Baffin Island is N40W following a directional that traces the northeast coast. A second directional spaced 5.6 deg farther west outlines the lower coast of the small peninsula on the southeast, and a third directional, again 5.6 deg to the west, traces the southwest coastline of the offset in the center of the island. Across another 5.6 deg interval is a fourth directional that lies along the southwest coast of the multilobed peninsula which hangs off the southern end of the island. Here again, as in so many other coastlines, there is a marked general directional trend combined with apparent offset shifts across intervals which approximate multiples of 2.8 deg.

On the right is a portion of Southeast Asia which includes many of the coastline examples shown before. On the northwest is Burma which connects with the Indochina Peninsula to the southeast and the Malay Peninsula leg hanging down to the south. Just below the Malay Peninsula across the Strait of Malacca is Sumatra in west Indonesia. East of Sumatra are the islands of Java and Borneo. The climate in the area is dominated by very distinct monsoonal wind patterns which mark the seasons. The large Asian land mass to the north is shut off from Southeast Asia by high mountains which hold back the air flow from that direction. During the arctic winter the air over Siberia grows intensely cold and pours out into the Pacific, moving south to bring the heavy northeast monsoon rains to Southeast Asia. The approach of the winter monsoon rains can be followed day by day as they move south down the Malay Peninsula and into Sumatra. When summer returns to northern Asia, winter comes to Australia. The cooler air moves north across Southeast Asia, bringing with it the southeast monsoons and summer thunderstorms which give Bogor, Indonesia more thunder days than any other place in the world. The general trend in the area is again N40W with clearly defined intervals between the coastline directionals. One interval of 11.2 deg lies between a directional along the northeast Indochina coast and another which lies along the Burma coastline. West across another 11.2 deg interval, another N40W directional traces the coast of Sumatra. Three more examples are given on the next page.

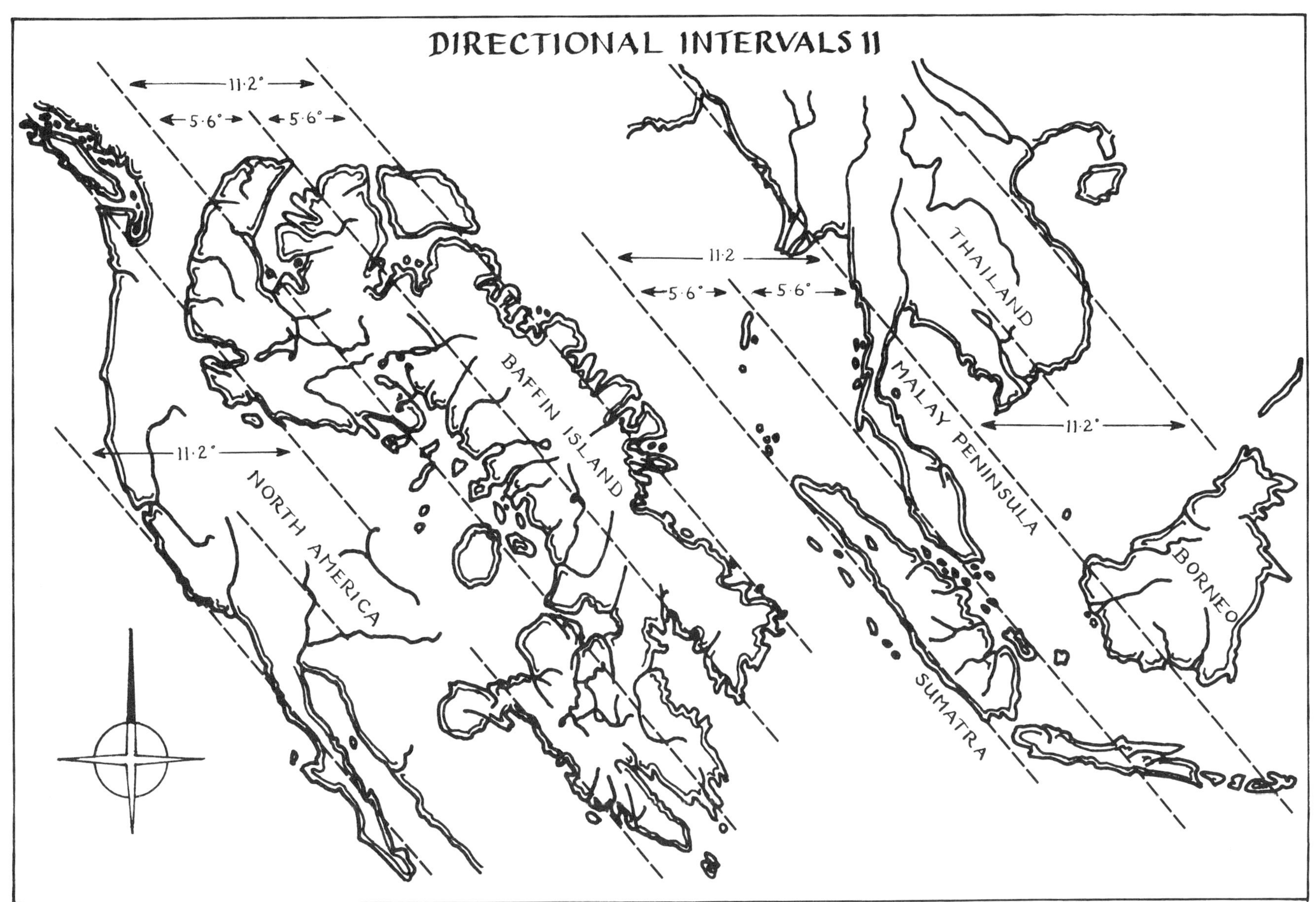
DIRECTIONAL INTERVALS 11
11·2°
5·6°
5·6°
11·2
5·6°
5·6°
11·2°
THAILAND
BAFFIN ISLAND
MALAY PENINSULA
NORTH AMERICA
11·2°
BORNEO
SUMATRA

1-9 Directional Intervals III

On the left of the figure is the west coast of Europe which borders the English Channel and the North Sea. The coastline runs northeast from Brittany in France along the coasts of Belgium, Holland, and Germany to Denmark, and then east along Poland to Lithuania and Latvia. Three N50E directionals are drawn which follow the general trend. The middle directional traces the linear continental coast of Europe from Brittany to Denmark. Another N50E directional across an interval of 5.6 deg to the east just touches the northern tip of Latvia, and a third to the west follows the northwest coast of Denmark. The word Denmark comes from the two Indo-European roots "dan," meaning low ground, and "merg," meaning border territory, giving a meaning something like "low border country." West across the North Sea are the British Isles which have many close historical ties to Denmark. The name England comes from the Angles of Denmark who crossed the North Sea and settled in Britain in the 5th century. The Angles were probably named after the shape of their angular homeland of southern Denmark. In the 9th century the Danes again came across the North Sea as Norsemen to conquer and rule England for a short time, and 200 years later in the 11th century England once more became part of the Danish Empire under King Canute. Meanwhile, another group of Danes had established the Duchy of Normandy in France, just across the channel from England, and in the famous year of 1066 the displaced Dane, William the Conqueror of Normandy, invaded England to become the first of the British monarchs after defeating Harold at Hastings.

The example in the center shows the coastlines of arctic northeast Russia. In the middle is the Okhotsk Sea bordered on the north and west by mainland Siberia, on the south by Sakhalin and the Kuril Islands, and on the east by the Kamchatka Peninsula. Kamchatka is directly in the path of the cold monsoon winter winds that blow from Siberia to Southeast Asia. As the winds cross the Okhotsk Sea, they soak up the moisture and then drop it over Kamchatka as extremely heavy snowfall. The peninsula lies on the Pacific Ring of fire and has frequent earthquake and volcanic activity. One of the largest eruptions in recent years was from the Volcano Bezymianny in 1956 which carried a huge mushroom cloud of ash to a height of 20 miles and filled a nearby valley with a thick layer of hot ash, creating the Kamchatka Valley of Ten Thousand Smokes similar to the one in Alaska of the same name.

Starting from Kamchatka and moving west, a number of N50E directionals have been drawn separated by 5.6 deg intervals. The first touches the tip of Kamchatka and traces the Kuril Islands southwest to Hokkaido. The next follows the coastline above Kamchatka down through the peninsula, and on to separate Hokkaido from Sakhalin. The third touches the north coast of Kamchatka and then carries on southwest to the Asian coast below Sakhalin. The fourth traces the northeast border of the Gulf of Selikhova in the far north, and extends southwest to follow the Amur River on the Asian mainland. The next cuts off the small offset on the south coast of the Okhotsk Sea, and the last follows the lower portion of the sea's west coastline bordering Siberia. About 1,500 miles further west, close to the Stoney Tunguska River on the Siberian Plateau, a huge explosion shook the Earth in 1908. Bright nights occurred all over Europe and huge displays of southern lights from a great electromagnetic storm were seen in the Antarctic. Although it was several years before an expedition reached the inaccessible area, it had hardly begun to recover from almost complete devastation within a 30 mile diameter circle. Something appeared to have exploded about five miles above the Earth's surface, leaving no crater or debris on the ground. Arguments of the possible cause went on for years, with explanations offered from an exploding space ship to a black hole penetrating the Earth. More recent study shows the culprit was probably a small piece of comet made of ice.

On the right is the east coast of North America which has been seen twice before. In the north is Labrador with Newfoundland off to the southeast. Just south is the mouth of the St. Lawrence River with Nova Scotia jutting out into the Atlantic. Further south is the highly industrialized east coast of the United States which runs along a central inset, and in the far south is the Carolina coastline. Four N50E directionals divide the coastline across equal 2.8 deg intervals. The first on the east touches the southeast tip of Newfoundland, and the next to the west parallels Nova Scotia and the Carolinas. The third directional splits off Newfoundland from the mainland and traces the central inset, while the last cuts through the mouth of the St. Lawrence River. Other directionals are drawn in that trace other features along the coast. On the next page both directionals and intervals will be brought together to form a pattern.

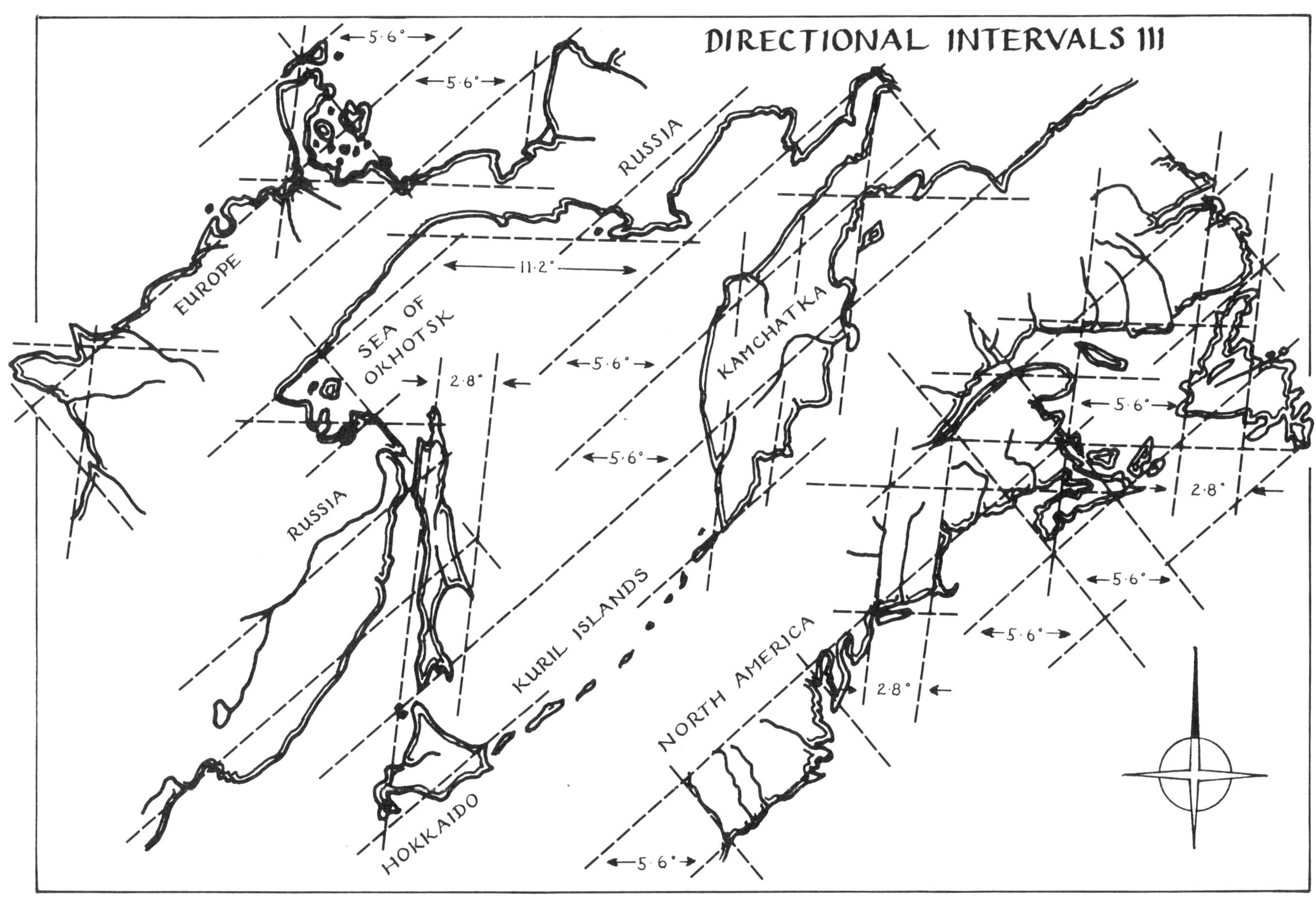
DIRECTIONAL INTERVALS III
5·6°
5·6°
RUSSIA
11·2°
EUROPE
SEA OF OKHOTSK
KAMCHATKA
2·8°
5·6°
5·6°
5·6°
RUSSIA
2·8°
5·6°
KURIL ISLANDS
5·6°
NORTH AMERICA
2·8°
HOKKAIDO
5·6°

1-10 Directional Intervals IV

Here again is the north central Mediterranean with a pattern overlay that includes all four directionals separated by 2.8 and 5.6 deg intervals. The pattern shows a repeating network of stacked rhombus figures crossed with diagonals. A rhombus is a four sided figure with opposite sides parallel and all four sides of equal length, resembling a squashed square with the top and bottom shifted in opposite directions. The word rhombus comes from the Indo-European root "wer" which means to turn or bend. In Greek, rhombus is the name for a magic wheel, or bullroarer, which makes a roaring sound when whirled around on a string. In the pattern, the sides of each rhombus are made of four directionals of equal length, with two N7E sides which are separated by 2.8 deg and which form the left and right uprights, and two W2N sides forming the top and bottom. The diagonals are lengths of N40W and N50E directionals which crisscross the rhombus from corner to corner. As mentioned above, the pattern is repeating and appears to be hierarchic, which means that it can be repeatedly split up into smaller and smaller figures of similar shape. By doubling the number of directionals, each rhombus is divided into four smaller rhombus figures one-quarter of the original size. The process can be repeated, giving a graded series of similar figures which decrease in area by one-quarter each time the number of directionals is doubled.

Four rhombus figures are shown on the illustration with one in the top row covering northern Italy, and three across the bottom row covering Corsica and Sardinia on the left, lower Italy in the center, and Greece on the right. On the left, the southern island of Sardinia is boxed in by two N7E directionals and two N50E directionals, one of which is a diagonal. Just north, Corsica is squeezed into the upper lefthand corner of a rhombus. Upper Italy is bracketed by two N40W directionals, one of which is a diagonal, and Sicily is separated from lower Italy by the bottom W2N side of a rhombus. This W2N side can be extended east to split off the Peloponnese Peninsula from mainland Greece which is bordered by two N40W directionals, one of which is a diagonal. Just northwest of Greece is the coast of Albania which lies on a N7E rhombus side, while the Turkish coast across the Aegean Sea to the east generally follows N40W.

For the ancient Mediterranean navigator directions meant winds, and he used a wind compass. In the west Mediterranean on the coast of Spain the strong "Levanter" blows from the northeast in the spring and autumn, and the hot, moist "Solano" from the southwest during the summer. Wind direction is always where it comes from, not where it goes. A sailor is far more concerned with what is coming at him, rather than what might happen some place else after the damage is done. On the south coast of France the cold, dry "Mistral," or Master Wind, from the north blows down the Rhone Valley and out into the Mediterranean, bringing with it heavy clouds, squalls, and rough seas. To the east, the hot, dry south wind from the North African desert called the "Scirocco" often blows across Sicily and lower Italy. In the Adriatic Sea between Italy and Yugoslavia a cold, violent northeast wind called the "Bora" rages in the winter, sometimes with hurricane force. The word "Bora" is Venetian for "Boreas," the blustery Greek god of the north wind. In contrast, the Adriatic summer is dominated by the light, fresh "Maestro" Wind from the northwest. Below the Adriatic in the Ionian Sea between Italy and Greece, the galeforce northeast "Gregale," or Greek wind, may blow for days, bringing with it cold, rainy weather. East of Greece in the Aegean Sea the predictable north "Etesian" winds come during the summer. "Etos" in Greek means year, but locally the winds are called "Meltemi" which is probably from the Venetian "bel tempo," meaning good weather.

In the Mediterranean the tides are small and hardly noticeable so that the changing winds are the sailor's main concern. Unlike many places in the world, winds control the Mediterranean surface currents which can change quickly if the wind suddenly shifts during a squall or thunderstorm. Our ancestors who sailed the Mediterranean and lived with these changing winds developed patterns to describe the characteristics of each wind and the interplay of geographical and seasonal relationships. Each wind was categorized by name and attributes, much as science today splits apart portions of nature into named groups of items. Stories of the changing winds probably took the form of relationships between different personalities who worked, talked, played, loved, and fought together. This would provide a method for the practical and needed knowledge of wind patterns to be passed down through the generations. To a degree, the stories might be called an explanation or model of the observed pattern. In the next chapter a possible explanation for the observed directional pattern will be developed after a brief review on the following page of this first chapter.

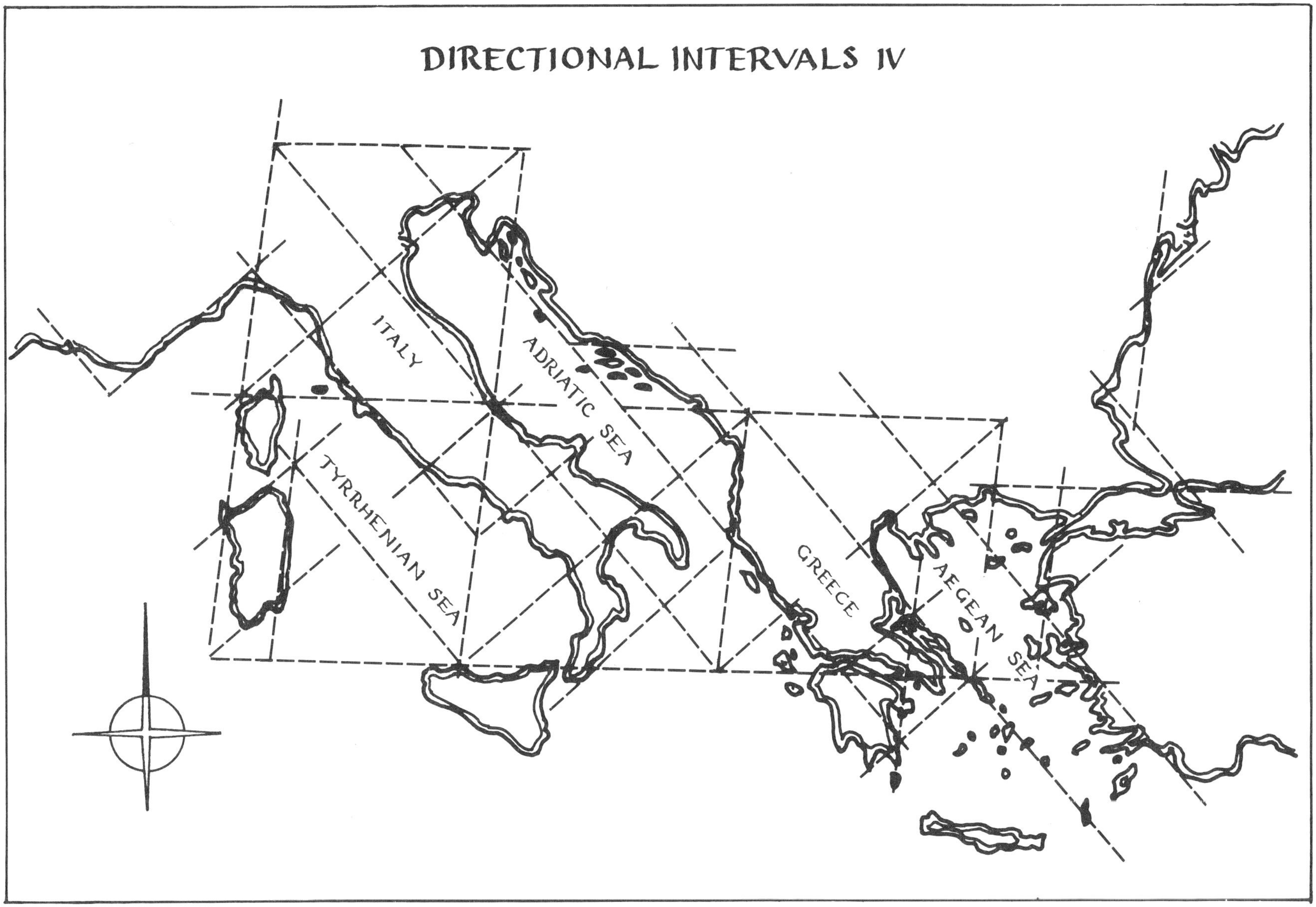

DIRECTIONAL INTERVALS IV
ITALY
ADRIATIC SEA
TYRRHENIAN SEA
GREECE
AEGEAN SEA

A number of examples have been given of relatively straight coastlines on Mercator Projection that follow one or more of four directionals. The first directional was defined using the west coast of North America that follows a linear trend of north 40 deg west, and the second, the east coast of North America that trends north 50 deg east. The directionals were correspondingly named N40W and N50E, and several examples were shown from a number of places in the world. It was pointed out that the two directionals intersect in a right angle. A third directional was defined using the long straight coast of South America that trends north 7 deg east. This directional, called N7E, was also found to be widely distributed around the Earth, although the trend was somewhat less clear than that of the first two directionals. The fourth and least clear directional was defined using the coastline of an arctic island north of Canada. It followed a west 2 deg north direction, and was accordingly called W2N. Several additional examples were given in which combinations of the directionals were used to trace coastline configurations. Similarities in the patterns included the box-like rectangular shape of the Iberian, Taymyr, and Indochina Peninsulas, the angular bends in the New Zealand and Japan Island groups, the triangular bulges jutting out from the coastlines of Alaska and Sweden, the many offsets such as those along the east coast of North America, and the stair step arrangement of alternating directionals running down through the Red Sea.

It was also observed that the alternating peninsulas and seas along the north Mediterranean coast seemed to be equally spaced. The equal intervals, as measured along a horizontal on Mercator Projection, were found to have a basic unit of 2.8 deg, or by doubling, 5.6 and 11.2 deg. Additional examples were given indicating that this equal spacing can be found in other parts of the world. The four directionals were brought together to form an equally spaced network of repeating rhombus figures crossed with diagonals. The N7E and W2N directionals traced the sides of the rhombus, and the N40W and N50E directionals the diagonals. If the number of lines in the network were doubled, cutting the intervals by half, a similar pattern would appear with four times the number of rhombus figures. This process could be repeated, indicating that the pattern may be hierarchical.

However, the observation of a possible worldwide hierarchical pattern of rhombus figures in selected examples is only a beginning. The observation requires the development of an adequate explanation relating the pattern to something other than examples. The explanation could take many forms from a sophisticated numerical model to a simple qualitative explanation based on analogy. The primary objective of explanation is to show that there is a reasonable, understandable mechanism that could be responsible for the observed pattern. Of course, the mechanism must be based on known and accepted physical principles. In our modern world of high technology this often takes the form of a computer model in which mathematical relationships determine the effect of combining certain physical variables. Setting up such a model requires that the overall process first be described and outlined in some detail to provide an organizational scheme for the many mathematical relationships that are required. This preliminary scheme is often called a conceptual model, and the explanation that follows in the next chapter is essentially such a model, concentrating on a simple qualitative analogy. A conceptual relationship of this nature provides the basis for the future development of an observed pattern from nature. New ideas issue from the explanation that more strictly define the role of the pattern in nature, and also bring up possibilities for extending the pattern over a broader base. Of course, this is a repeating process in which selected examples bring out a pattern that leads to an explanation, that in turn leads to other examples that are used to adjust and improve the pattern. Although this repeating process of pattern improvement is never complete within the indi-

visible totality of nature, each step increases the scope of understanding within a bounded and constrained framework.

Before proceeding to an explanation, it is again emphasized that the examples used are selective, and that there are many coastlines that do not conform to the directionals. This raises two important concepts related to geologic analysis. The first is the necessity for a multiple approach to most questions in geology. Nature can be looked at in many different ways, and more often than not new patterns are merely new supplementary tools to be added to those already in use. The second concept is the requirement that an explanation include not only the closely matched examples, but also those aspects of nature that do not seem to conform. The major inconsistencies and anomalies of the directional pattern will be discussed from from time to time as they appear. However, the next chapter will concentrate on the more positive side of the explanation.

2 THE ANALYSIS

EXPLAINING THE PATTERN

As discussed in the first chapter, geology differs in degree from some of the other physical sciences. Geologic information comes primarily from on, or just below, the surface of the Earth. The enormous interior of the Earth, with its intricate and complicated mix of materials in an extreme environment, has never been seen or touched, but only crudely probed with waves from near surface shocks. However, the geologic patterns which are observed on the surface are generated by gravitational, chemical, magnetic, and electrical forces from within that huge unknown interior. These forces can rarely be duplicated in a laboratory because they are so large compared to those at the surface, and because they are acting on enormous complicated masses of material over periods of time which may be measured in millions of years. This results in a great deal of uncertainty and guesswork in attempting to scale up laboratory experiments to simulate the processes at work within the Earth's interior. The problem is further complicated by the uncertainty in the composition and structure of the masses of material involved in the process. To a degree, geologic analysis represents an attempt to understand an observed surface pattern generated by large uncertain forces acting on large uncertain masses of material over long uncertain periods of time.

The frustrating uncertainties of geology are also a factor in the individual understanding of just what the basic objectives of analysis might be. It is commonly understood that the end objective of science is prediction of the future, and that other aspects are only steps along the way. This concept of science is an ingrained corollary of the modern world. The daily weather report is a prediction from meteorologists of the chance of rain tomorrow. Popular articles frequently discuss the prediction of when scientists will be able to predict earthquakes. In recent years there has been a flood of predictions relating to the increase of carbon dioxide, the decrease in ozone, the rise in sea level, the melting of ice caps, and when the world will die. However, prediction in science is no better than the available uncertain information allows it to be, and the vague nature of much geologic information may preclude prediction, except in a very generalized sense. This forces the emphasis in geologic analysis towards a description and explanation of the present rather than a prediction of the future, and often leads to a misunderstanding between a geologist and a co-worker whose background is in the physical sciences. The geologist tends to consider his work virtually complete when he has adequately described the pattern under consideration and found a reasonable explanation for its existence, while the other party may feel that an analysis only begins at that point. Considering the uncertainty of available information, it may well be that a geologic explanation might best be limited to a qualitative understanding by analogy, and that prediction be restricted to general comments.

Analogy can be defined as a relationship between two patterns or processes which are similar in some respects. Each similarity in the analogy must be defined, and care must be taken not to imply that two things alike in some respects must be alike in other respects. However, if it is properly qualified, an analogy may provide the best explanation possible for a particular pattern. Although an analogy is not an end in itself, it does provide the framework of something familiar on which to organize the unfamiliar, and that in itself brings a degree of rationality to what might appear to be irrational. In addition, tracing possible similarities may open up new directions for testing other aspects of the original pattern.

In this chapter Mercator Projection will first be defined, explaining why a straight line on Mercator Projection represents a constant compass direction on the globe. As discussed in the previous chapter, the four directionals are straight lines on Mercator Projection which correspond to constant compass directions on the globe. The flat Mercator map will then be divided into a network of unit squares, and it will be shown that if the earth is twisted around its poles, the Mercator unit square network will be deformed into a pattern of rhombus figures. This represents a mechanical analogy called pure shear in which a twisting force called "torsional stress" acts on a thin-walled cylinder. In this analogy the unit square is stretched into a rhombus with peak deformation along the diagonals, one of which is lengthened under tension and the other shortened under compression. It will also be shown that the rhombus pattern might be hierarchical, with the diagonal intersections acting as focal points for doubling the lines in the pattern. A hierarchical series, starting from the full 360 deg circumference of the Earth and repeatedly halved, would contain 11.2, 5.6, and 2.8 deg as members. This hierarchical rhombus pattern will then be shown to fit various Earth features including coastlines, ocean ridges, and ocean trenches. Aspects of the analogy will then be used to suggest possible ways in which Earth features might reflect the directional pattern. These suggestions will then be tested by example in succeeding chapters.

2-1 Mercator Projection

This figure shows the development of Mercator Projection, in which the round Earth is transformed into a flat map with straight lines representing constant compass lines on the globe. On the left is the Earth with its surface marked off with two sets of imaginary circles called meridians of longitude and parallels of latitude. The meridians are great circles running north and south which intersect at the poles, dividing the surface into crescent shaped slices. The meridian running through Greenwich, England is defined as zero, and degrees are measured east and west of this line. It is perhaps easier to think of the equator as divided up into 360 equal arcs, each representing one degree of the full 360 deg in the equator circle. These degree arcs are marked off by 360 great circle meridians running north and south which cross the equator and meet at the poles. The separation between meridians is always one degree, although the true global distance varies from a maximum at the equator to zero at the poles where they meet.

The parallels of latitude are circles on the globe parallel to the equator which get smaller and smaller as the poles are approached. Consider the Greenwich or zero meridian great circle which crosses the equator and passes through the poles. It can be divided into 360 equal parts, representing the 360 deg circumference, and circles parallel to the equator can then be drawn on the globe marking these divisions. This gives 90 circles above the equator at one degree intervals, which decrease in size until the 90th circle at the North Pole is only a point. There are also 90 circles below the equator, with the 90th circle a point at the south pole. For these parallels of latitude, degrees are measured either north or south of the equator. Any place on the globe can be located using these two imaginary sets of circles.

Mercator Projection is called conformal, which means that the shape of a small area on the map is similar, or conforms, to its shape on the globe. This is the same as stating that all angles around any point on the globe are correct on the map. The word conformal is from Latin, and means the same or similar form. Sometimes the word orthomorphic from Greek is used, meaning correct form. Although small areas are true to form, large areas are severely distorted, with those near the poles, such as Alaska or Greenland, appearing much larger on Mercator Projection than they truly are on the globe. In addition, the projection has the very special property that any straight line on the map represents a constant direction on the globe. A navigator on a ship can draw a straight line between any two points on a Mercator Map, measure the direction angle, and then set a steady course with that direction as a constant bearing. This special property is why the projection is so widely used in navigational mapping. The name Mercator is from Gerardus Mercator who developed the projection and published the first Mercator map in 1569. He was born in Belgium as Gerhard Kramer, but later Latinized his name which was common practice at that time. At the university he studied under the famous mathematician Gemma Frisus, and then started his own business of scientific instrument making and map engraving. After trouble with the Inquisition, he moved to Duisburg on the Rhine and eventually became well-known over all of Europe as a geographical consultant. It seems probable that he developed his Mercator Projection empirically, and left the mathematics for others to work out.

As shown on the left, Mercator Projection can be visualized by taking a cylinder and sliding it over the Earth so they touch all around the equator. A point on the Earth's surface is then projected on to the cylinder by sighting a straight line from the Earth's polar axis through the point on the Earth, and then on to the surface of the cylinder. After all points and lines on the Earth have been projected onto the cylinder, it can be unrolled to make a flat map. It is evident that a unit area on the globe would increase in size on Mercator Projection as the poles are approached, and that the poles themselves can never appear on Mercator Projection, because the line joining the poles is the axis of both the globe and the imaginary Mercator cylinder.

The map shown on the right is the unrolled cylinder which contains a Mercator Projection map of the world. This map was traced from Figure 10.5 in "Physics of the Earth" by Frank D. Stacey, 1977. The solid lines represent coastlines enclosing land areas. The sawtooth lines follow trends on which many volcanoes and earthquakes can be located, and which generally trace the deep ocean trenches. These deep gashes in the Earth total some 16,000 miles, and are concentrated around the Pacific Ocean where they follow the earthquake and volcano prone Ring of Fire. The dotted lines generally follow the mid-ocean ridges which stretch for more than 19,000 miles at an average height of 8,000 feet above the ocean floor. A continuous ocean ridge circles Antarctica, isolating it from the continents to the north, and other ridges branch off northward to split the Atlantic, Indian, and Pacific Oceans. Portions of dotted and sawtooth lines can also be seen onshore where evidence indicates effects similar to those observed offshore. In the pages that follow, the Mercator map itself will not be changed, but a new grid of imaginary sets of lines will be developed in which the meridians are retained and the parallels adjusted.

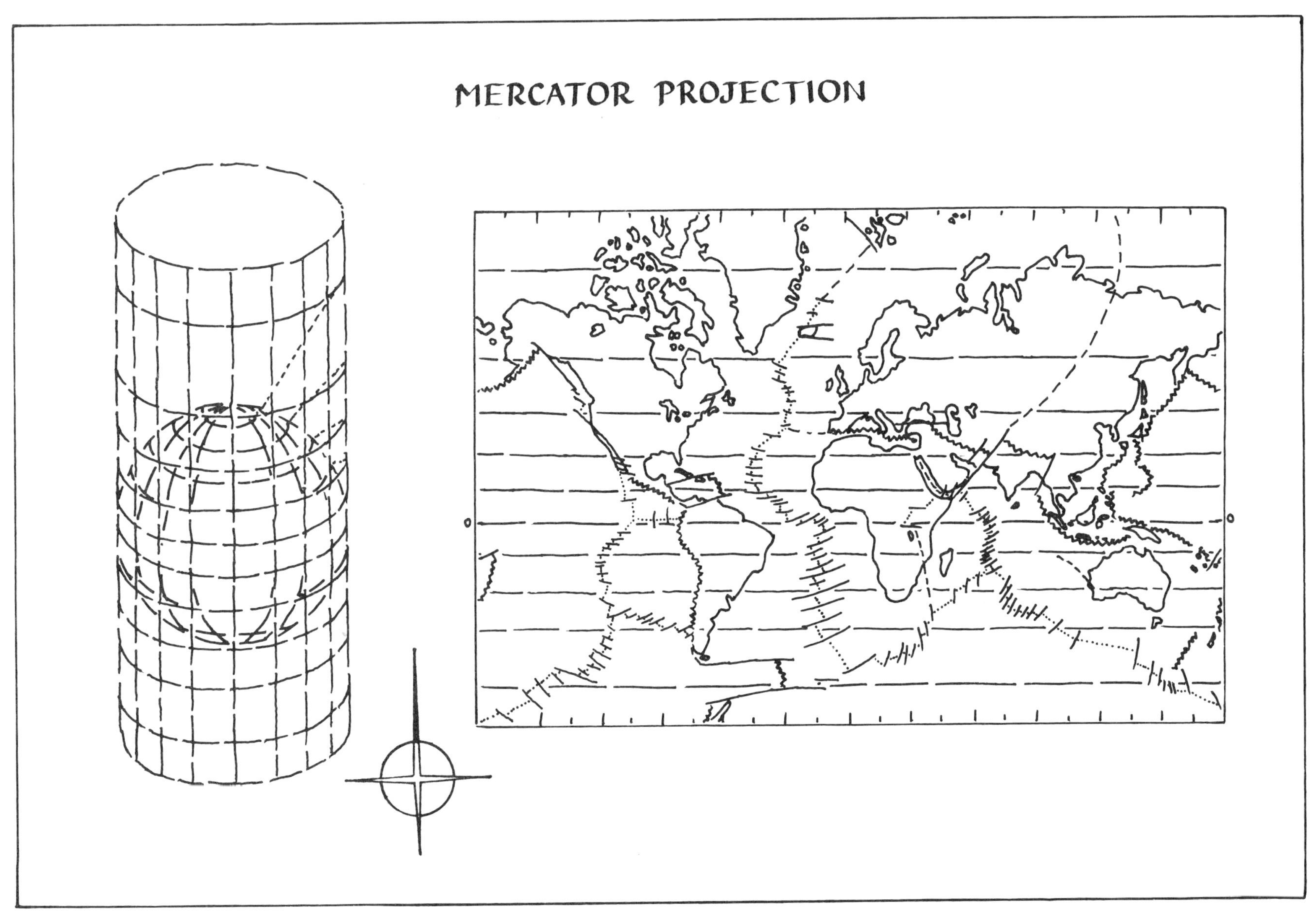

MERCATOR PROJECTION

2-2 Unit Square Cylinder

In this figure the Earth globe has been taken out, leaving only the cylinder and the rolled out Mercator Projection. On the previous figure the parallels of latitude were shown on the map together with border markers indicating the meridians of longitude. It will be remembered that these meridians are equally spaced, parallel lines running north to south, and that the separation between any two meridians is always the same number of degrees no matter how far north or south of the equator. However, the parallels of latitude are not equally spaced on Mercator Projection. It will be recalled that they represent circles parallel to the equator which grow smaller as the poles are approached. They are equally spaced on the globe along a north to south great circle running through the poles, but when projected on a Mercator map, they form a set of parallel lines running east to west which are spaced further and further apart as the poles are approached. These two sets of meridians and parallels are only imaginary, and are defined as a convenience to locate places on the globe. If properly defined, other imaginary sets of lines might just as well be determined. Depending on the way they are to be used, adjustments of this type can often simplify explanations.

The rolled out Mercator map on the right shows a grid of two line sets. The 16 equally spaced parallel lines running north to south are meridians of longitude spaced at intervals of 22.5 deg. The number 16 was chosen because it represents the fourth member of a doubling series which starts with 2 and runs 2, 4, 8, and 16. This is consistent with the observations from the previous chapter which indicated the possibility of a hierarchical pattern based on a doubling series. The second set of equally spaced parallel lines runs east to west spaced at intervals equivalent to those of the first set. The two sets of lines generate a network of repeating squares, each of which will be called a unit square. Working backwards, as compared to the discussion on the previous page, the flat Mercator Map with its unit square grid can be rolled up into the cylinder on the left, which will then be called a unit square cylinder. A brief review of the Earth's interior provides the necessary background for the analogy.

The Earth is a sphere about 7,900 miles in diameter with a slight flattening at the poles due to the rotation. The flattening is very small, with a difference of only 13 miles between the radius at the pole and at the equator. There is also a very small additional flattening of 130 feet at the South Pole which makes the Earth slightly pear-shaped. The Earth has a thin skin between 4 and 30 miles thick called the crust with a volume less than one percent that of the total Earth. This is the only portion of the Earth that can be directly observed and measured. The thin crust is relatively hard and brittle, and has a tendency to fracture or break after only slight deformation. Recent evidence indicates that the crust is variable in composition and structure, and with many layers, especially in the lower parts, that have a tendency to slip, slide, and deform. Just below the thin crust is the thick mantle which contains about 80 percent of the Earth's volume. The mantle is solid, but tends to deform by flow rather than fracture because of the high temperatures and pressures. This plastic flow, or creep, is a very slow process taking place over millions of years. The difference between the thin hard crust and the thick soft mantle allows the crust to act somewhat independently of the mantle. Below the mantle is the 2,150 mile thick core, divided into a 1,300 mile thick liquid outer core and a 850 mile thick solid inner core at the center of the Earth, probably made up mostly of iron. The magnetism of the Earth may be a result of liquid circulation in the outer core which generates electromagnetic currents in the inner core of solid iron.

This structure of the Earth's interior described above is primarily based on the study of seismic waves, the vibrations that follow natural or man-made earthquakes. There are several kinds of vibrations, or waves, with different characteristics. One type is an up and down rolling motion like the waves made from a stone dropped into still water, and another type relates to a push and pull cycle of compression and relaxation in material. Other vibrations shake the entire surface of the Earth much like a ringing bell. Each wave type has a different velocity depending on the composition of the material involved, so that the composition of the Earth's interior may be estimated by comparing the travel time of the various wave types. Although the estimates are of a broad nature, improvements in techniques and instrumentation have added much detail in recent years. However, there will always be questions due to the improbability of ever collecting samples from below the Earth's crust.

Although questions remain, there is a major item of agreement, and that is the difference between the thin surface crust and mantel underneath. The crust is a relatively hard thin shell covering the Earth and enclosing the soft plastic mantle just below. As a first approach, the Earth would probably behave as a thin-shelled sphere if a force or stress were applied. The resulting strain, or deformation, in the thin shell from a twisting force will be discussed on the next page.

UNIT SQUARE CYLINDER

2-3 *Twisted Cylinder*

This figure is the same as that on the last page, except that the cylinder on the left has been twisted. A twisting stress, or force, of this kind is called torsion and is comparable to wringing out a towel. The cylinder is formed of a thin sheet of material rolled up to make a thin-walled pipelike cylinder, and when a torsion force is applied the process is called pure shear. It is assumed that the cylinder keeps its circular shape, but that the material in the thin wall adjusts to the twisting force by stretching in some places and contracting in others, much like a thin layer of rubber. The twisted cylinder is unrolled and laid out on the right to show how the unit squares have been deformed. Although opposite sides remain equal and parallel, the top of each unit square has been shifted to the right and the bottom to the left, changing the unit square into a rhombus. The study of how materials deform when a force is applied is a branch of mechanics in which the applied force is called stress and the resulting change in shape is called strain. These words are commonly used in much the same way when speaking about people. The stress, or pressure, of a job can sometimes cause strain that shows up as a change in personality characteristics.

There are many forces acting on the Earth which could be reflected in a twisting stress. The rotation of the Earth tends to pull on material around the equator causing an equatorial bulge and a slight flattening at the poles. Differences in the distribution of material within the Earth could introduce a difference between the rotational forces at the North and South Poles, giving a slight twist to the Earth. There are also gravitational forces acting on the Earth from the nearby Moon that pull up the oceans into a wave that follows the daily circle of the Moon around the Earth, creating the tides. Mutual gravitational forces also act between the huge rotating Sun and the Earth, that circles the Sun in a plane called the ecliptic. The rotational axis of the Earth is tilted relative to the plane of the ecliptic, so that the gravitational force acting on the Earth is offset from the equator. This forces a turning motion of the Earth's axis called precession, which traces out every 25,700 years relative to the stars an hourglass figure with the thin waist located at the center of the Earth. Precession is reflected in the slow movement of the heavenly North Pole through the pattern of stars. The North Pole is now located close to Polaris, the North Star, but some 9,000 years ago it was in the constellation of Hercules, who knelt above the spinning Earth with his right knee on the North Pole of the sky. The Earth also has many small periodic anomalies which are called nutations, wobbles, or l.o.d., meaning length of day changes. These irregularities are probably related principally to the uneven distribution of the Earth's internal composition.

There is another twisting force which has only recently begun to be understood. The Earth acts as an enormous magnet, with the magnetic field looping out like a huge doughnut from the magnetic poles which are located close to the rotational poles. This magnetic field is probably generated within the Earth's core as the molten iron of the outer core slowly rotates and circulates. As with all magnetic fields, there is a related electrical field which combines to form an electromagnetic field. The Sun also has an electromagnetic field, but with complications. Indications are that the electromagnetic field lines of the Sun are twisted and contorted, and that enormous bursts of energy are released which appear as sunspots or flares when the lines break out through the surface. There is also a continual outflow of ions and electrified particles, called the solar wind, which spews out from the sun in concentric spirals like a water sprinkler. The electrical charge of the solar wind alternately shifts between plus and minus for each quarter of the Sun's 27 day rotational cycle. The solar wind is very gusty, with large variations in both the velocity and quantity of the material that comes from breaks in the Sun's surface called coronal openings. This huge warped spiral of charged material is tilted by 7.2 deg relative to Earth's orbit, due to the 7.2 deg inclination of the Sun to the ecliptic.

As this tilted, gusty pinwheel of charged particles streams out from the Sun, it drags along the Sun's magnetic field lines. As the Earth is approached, the solar lines are pushed away by the Earth's magnetic field so that they curl around the Earth and stream out behind. The interaction between the solar wind and Earth releases large amounts of energy observed as auroras, often called northern or southern lights. There is still much uncertainly about these interactions due to the limited information available. Space probes have all traveled in the plane of the ecliptic, so that a three dimensional picture including the effect of the 7.2 deg tilt is very speculative. Additional information will come from the Solar Polar Mission which will fly over the Sun's poles. Certainly, the twisting force of the solar wind would have some minor effect on the Earth's rotation. Any or all the forces mentioned above could be responsible for a slight torsion or twisting force around the axis of the Earth. However, the specific combination of forces that could be responsible is not immediately obvious.

TWISTED CYLINDER

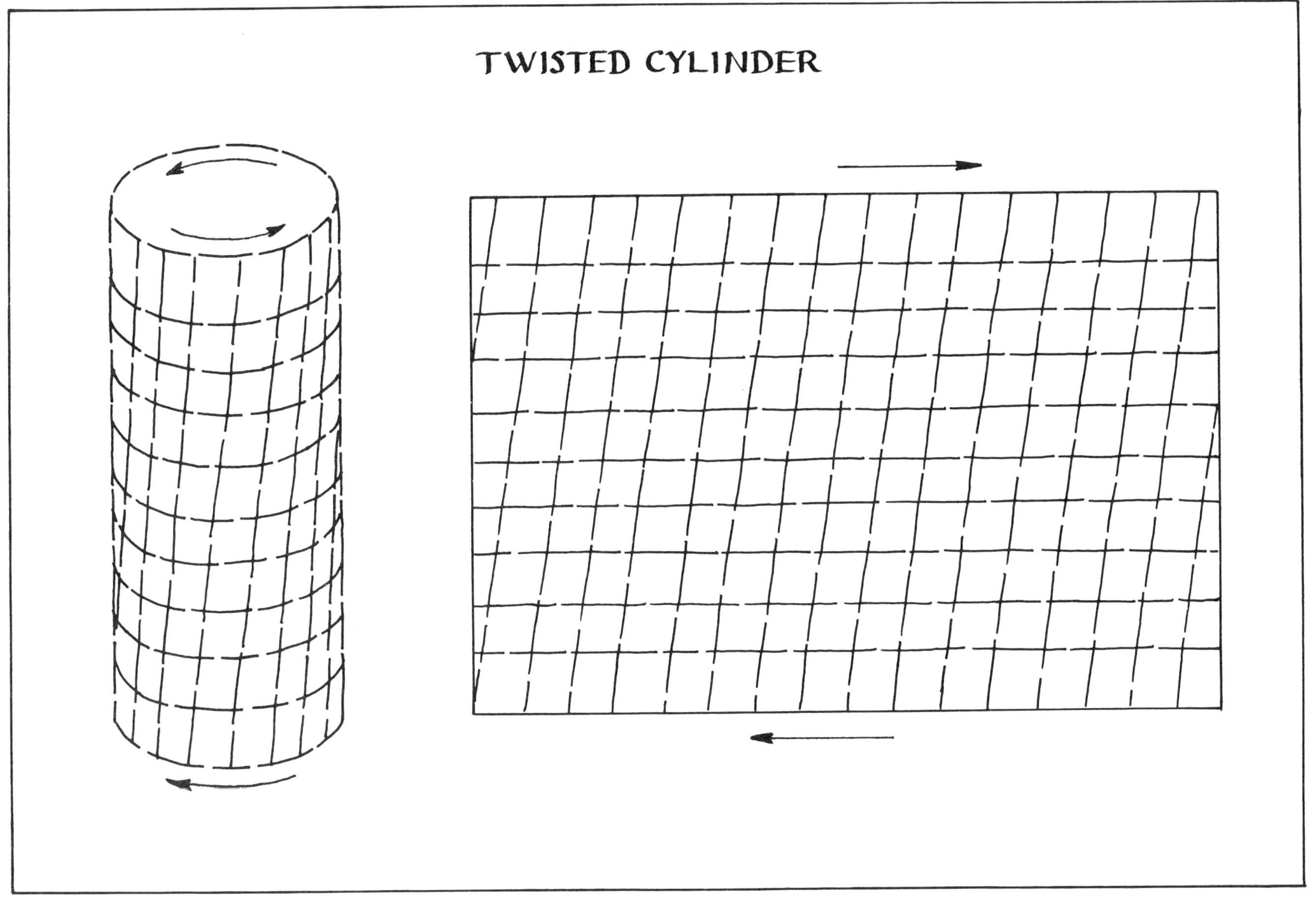

2-4 Twist Deformations

This gives a more detailed look at the shape of a unit square on a thin-walled cylinder after a pure shear, twisting stress has been applied. The original unit square is shown in the upper left, and just to the right is the changed rhombus shape after twisting. The peak deformation, or strain, is along the rhombus' diagonals which always intersect in 90 deg right angles. The diagonal from lower left to upper right is lengthened and the diagonal from upper left to lower right is shortened, and both diagonals have been rotated 5 deg clockwise. Although the sides of the unit square are unchanged in length, they have been deformed by tilting. If the uprights are tilted clockwise by 7 deg and the horizontals by 2 deg, the rhombus figure is identical to the observed pattern shown in the last chapter. This close correspondence between the strain rhombus and the observed rhombus is reinforced by other similarities. As mentioned above, the diagonals of the strain rhombus, which represent the N40W and N50E directionals, are lines of peak deformation, and it was observed that these two directionals are most clearly observed on the globe. The N7E directional was observed to be somewhat less clear, corresponding to the uprights on the strain rhombus which are less deformed than the diagonals. Least clear of all the directionals was the W2N which corresponds to the least deformed top and bottom sides of the strain rhombus.

It was also observed that the four directionals were combined in equally spaced multiples to form what appeared to be a hierarchical network of rhombus figures. This feature also corresponds closely to the pure shear strain pattern. The strain pattern of the rhombus is considerably more complex than shown on the simple diagram. The intersection of the diagonals is the center point of a pinwheel pattern in which the strain gradually increases to an extreme high, then decreases to an extreme low four times around the pinwheel. The pattern might be thought of in terms of a clock with a wavy face, and the minute hand pivoting around the intersection point of the diagonals. As the clock hand sweeps out a full hour, the strain gradually increases to a wave peak and then decreases to a wave trough four times. The hand drops to least strain at one minute (N7E), 15.5 minutes (W2N), 31 minutes (N7E), and 45.5 minutes (W2N), and rises to peak strain at 8 minutes (N50E), 23 minutes (N40W), 38 minutes (N50E), and 53 minutes (N40W). In addition to this pinwheel pattern, the strain also varies along the clock hand, increasing outward from the diagonal intersection pivot point to a maximum at the corners of the rhombus.

The overall strain pattern has least strain lines which follow N7E and W2N and cross at the center pivot, which is a node of least strain. These least strain lines divide the sides of the rhombus figure in half, creating four smaller but similar rhombus figures. Assuming each smaller figure has similar characteristics, the process can be repeated a number of times, splitting each side segment in half over and over. This sequence is shown on the lower left, where the left and top side have been repeatedly cut in half several times to generate the linear series 1, 1/2, 1/4, 1/8, and so on. A pattern of this kind which can be repeatedly subdivided into smaller multiples of similar shape is called hierarchical.

On the lower center of the page a rhombus network is shown which was formed by dividing a rhombus side in half three times, to give 8 figures along a side for a total of 64 figures in the pattern. The tilt of the sides has been constrained by the hierarchical sequence to give north 7.2 deg east, the inverse tangent of 1/8, and west 1.8 deg north, the inverse tangent of 1/32, which closely match the observed values of N7E and W2N. The computed diagonals are north 40.5 deg west and north 49.5 deg east, which again closely match the remaining two directionals N40W and N50E. The correspondence can be further extended as shown by the sequence on the right. The equator circles the Earth midway between the poles, and is divided into degrees so that the total length around the Earth is 360 deg, the number of degrees in a circle. In Mercator Projection the 360 deg length of the equator is represented by the width of the rolled out cylinder, which corresponds to the unit side of the pure shear strain rhombus. The sequence is formed by repeatedly dividing the 360 deg unit side in half to create the hierarchical series beginning 360, 180, 90, 45, 22.5, 11.2, and 5.6 deg. These last three numbers correspond to the observed directional intervals shown in the last chapter, further reinforcing the correspondence between the observed directional pattern and the pure sheer analogy. From now on throughout the book, all directional patterns used will be based on the hierarchical sequence given on this page.

It has been shown that the strain pattern from pure stress is similar in many respects to the observed directional pattern shown in the first chapter. On the following page is the first example that explores this relationship. It shows an overall view of the entire world in Mercator Projection. The hierarchical strain pattern will be fitted over the map to show the correspondence between the pattern and the larger scale Earth features.

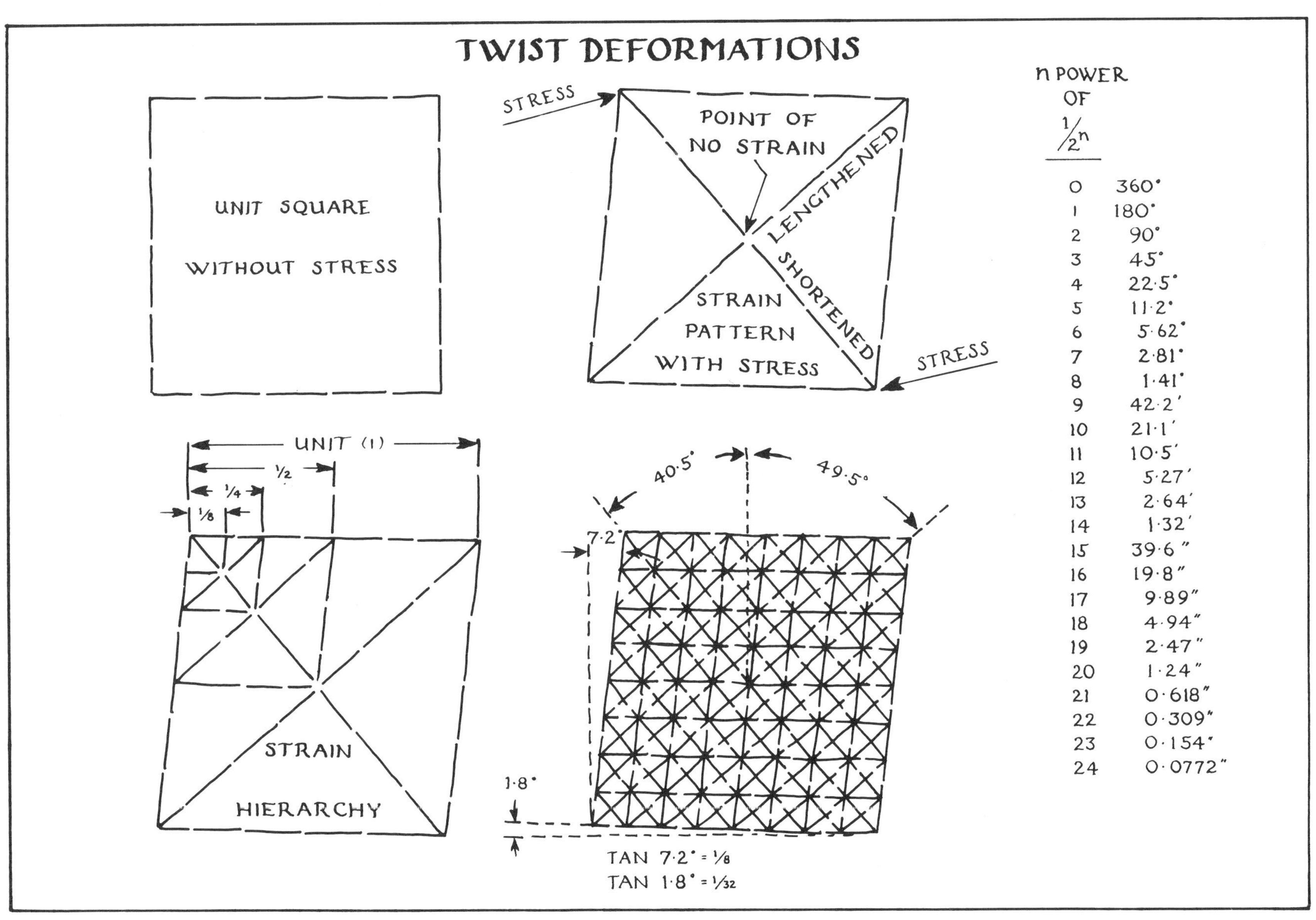
TWIST DEFORMATIONS
UNIT SQUARE
WITHOUT STRESS
STRESS
POINT OF NO STRAIN
LENGTHENED
SHORTENED
STRAIN PATTERN WITH STRESS
STRESS
n POWER OF 1/2^n
0   360°
1   180°
2   90°
3   45°
4   22.5°
5   11.2°
6   5.62°
7   2.81°
8   1.41°
9   42.2'
10   21.1'
11   10.5'
12   5.27'
13   2.64'
14   1.32'
15   39.6"
16   19.8"
17   9.89"
18   4.94"
19   2.47"
20   1.24"
21   0.618"
22   0.309"
23   0.154"
24   0.0772"
UNIT (1)
1/2
1/4
1/8
STRAIN
HIERARCHY
40.5°
49.5°
7.2°
1.8°
TAN 7.2° = 1/8
TAN 1.8° = 1/32

2-5 Global Directionals

On the left is the Earth with a Mercator Projection cylinder fitted around the globe and touching the equator. The network of unit squares on the cylinder has been deformed by twisting the cylinder through an angle of 7.2 deg, and tilting the circular lines by 1.8 deg so that they spiral around the cylinder as one continuous curve. The unrolled cylinder with its strain rhombus pattern, including both sides and diagonals, is shown on the right with a Mercator Projection map overlay of the world. The tilted verticals corresponding to the rhombus uprights are N7E directionals, and the tilted horizontals which correspond to the rhombus top and bottom sides are W2N directionals. The diagonals of the rhombus figures which crisscross the map and intersect at right angles are N40N and N50E directionals. The horizontal interval separating the directionals is 22.5 deg, equivalent to the fifth term of the hierarchical sequence on the preceding page. This strain rhombus grid is relatively easy to construct because it is based on the sequence of repeatedly dividing by two on a squared sheet of paper. The N7E directional is defined by counting eight squares up and one to the right, and the W2N directional thirty-two squares right and one down. The repeating rhombus grid can then be drawn using an interval from the hierarchical sequence, and finally the N40W and N50E directional diagonals drawn in through the rhombus corners.

On the Mercator Projection Map coastlines are shown as solid lines. The dotted lines, broken by short solid lines, show the mid-ocean ridges which are interpreted in Plate Tectonics to be lines of divergence where material from the mantle is oozing out to form new ocean floor. The short solid lines are the transform faults which cross the ocean ridges at intervals. The sawtooth lines on the map represent the deep ocean trenches and counterparts on land. In plate tectonics these trenches are interpreted as sub-duction zones where the crust of the Earth is broken,

and one side is sliding down under the other and into the mantle. On the Asian continent are two continental sawtooth lines, one following the Great Himalayan Range just north of India, and the other following the Elburz Mountains south of the Caspian Sea, with extensions west to the Black Sea and east to India.

Only selected portions of the strain rhombus grid are shown on the map to emphasize the correspondence between the strain pattern and Earth features. The general trends of continental coastlines are fairly well established by the directional grid lines. The North American continent is enclosed by an inverted triangle with the apex pointing downward into Mexico. The base of the inverted triangle follows a W2N directional running across the top of the continent. The two sides of the triangle follow a N40W directional on the west and a N50E directional on the east. The northeast corner of the triangle is clipped off by a N40W directional running along Labrador and Baffin Island, with another N40W just 22.5 deg to the east separating Baffin Island from Greenland. The Eurasian continent is outlined on the west by a N50E directional which follows the European coastline, and on the east by a parallel N50E which follows the Asian coast. South of Europe a N2W runs from Spain through the Mediterranean to the Caspian Sea where it meets a N40W which traces the west side of an inverted triangle enclosing India. The northwest bulge of South America is bordered by two N40W directionals, and the narrow leg in the south by two N7E directionals. Although the African boundaries are considerably more complicated, Australia is fairly well defined with two sets of rhombus diagonals forming the tip of an arrow pointing north.

Another feature of the correspondence are the offsets which are evident in a number of places. There is a 22.5 deg offset on the west coast of North America, and another in the N50E directional which traces the west coastline of the Eurasian Continent. The east coasts of both Asia and North America have a num-

ber of offsets, and west of Europe the British Isles, Scandanavia, and the Russian Novaya Island lie within a 22.5 deg interval between two N50E directionals. Many other offsets on a smaller scale can be found on the map.

A number of directionals can be followed for long distances across the map. The N40W which traces the California coastline can be extended southeast along the coast of Peru and the northern part of the Peru-Chile Trench, and on to the South Sandwich Trench just north of Antarctica. Just off the east coast of Asia, a N50E directional traces the Pacific Trenches, and can be extended southwest to follow the Southwest Indian Ocean Ridge just south of Africa. Another N50E directional follows the Baikal Lake Rift southwest across Russia, then on across Arabia to connect with the Afar Triangle in the Red Sea and the East African Rift. It is also interesting to note the close correspondence of the directionals with the mid-ocean ridges and trenches over much of the globe. It is important to remember that the directionals are straight only on Mercator Projection. In the next figure these constant direction lines will be shown as viewed on the true spherical Earth.

GLOBAL DIRECTIONALS

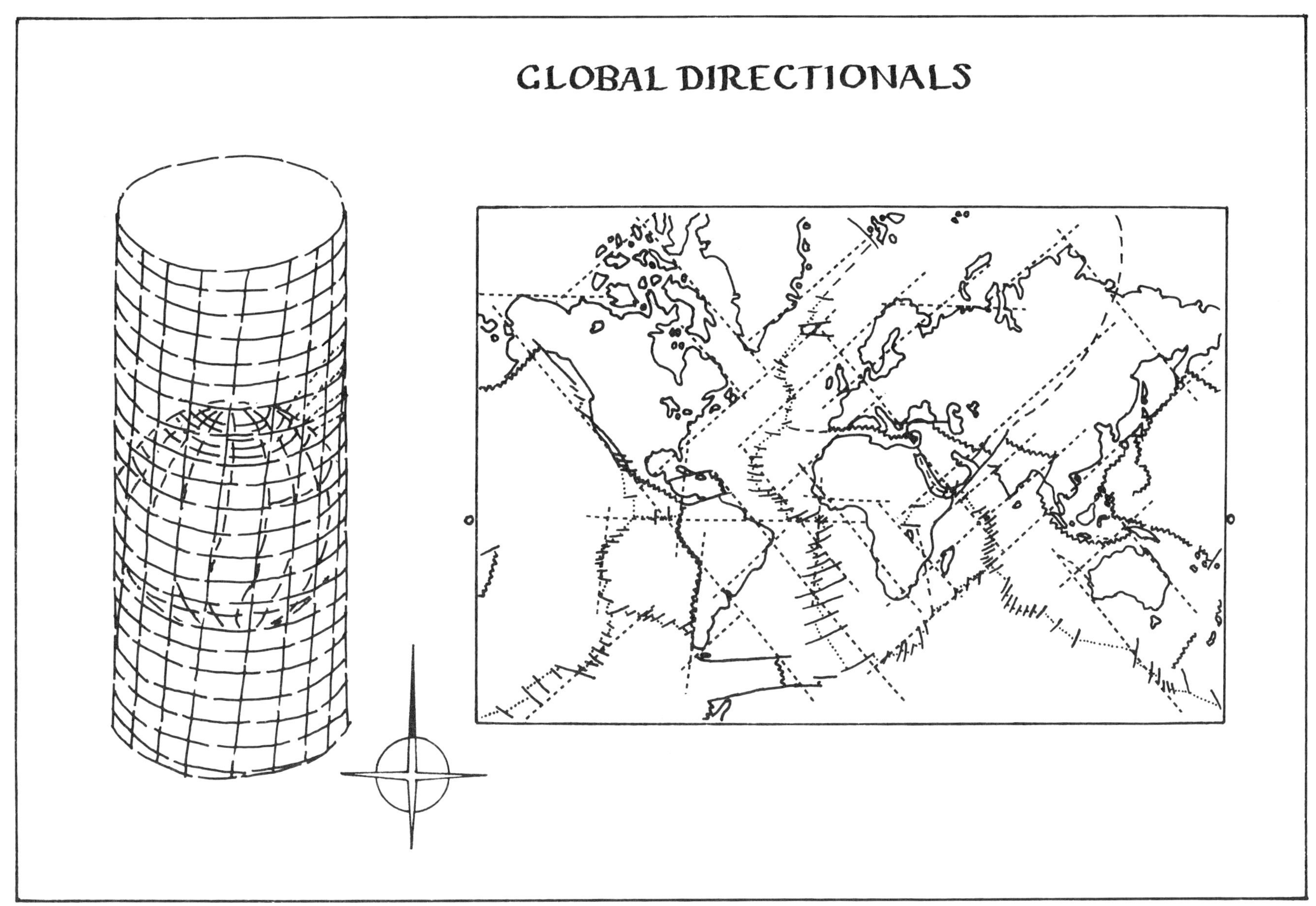

2-6 *The Twisted Earth*

This shows the spherical Earth as viewed from space. The two arrows at the top and bottom represent the direction of the twisting stress that causes the rhombus pattern strain over the surface of the Earth. Looking down from above each pole, the force acts in a counterclockwise direction. The four sided dashed line figure represents the Mercator Projection unit square projected back on the spherical Earth. The solid curved lines show the strain rhombus sides and diagonals after twisting, corresponding to the four observed directionals. Each maintains a constant direction relative to the poles as it spirals around the globe. The elongated S-curves running approximately north and south are N7E directionals which coincide with the rhombus uprights. The lines which appear to run east and west are actually one continuous N2W directional which spirals around the Earth, shifting up 1/32 of the 360 deg circumference each time it completes a circle. This N2W spiral coincides with the top and bottom sides of the strain rhombus. The diagonal N40W and N50E directionals run through the corners of the strain rhombus and gently curve across the surface of the Earth like the streaks in a round sourball candy. None of the directionals ever reach the poles, but rather spiral tighter and tighter together as the poles are approached. It is remembered that the Earth's axis, which passes through the poles, is the same as the axis of the Mercator Projection cylinder, so that the poles can never be mapped onto the cylinder surface. In the practical world of nature this would probably be observed as a gradually increasing degree of confusion and distortion in the directional pattern as the poles are approached, until the pattern was lost completely in the near vicinity of the poles.

These constant bearing directionals which spiral around the Earth approximate great circles over limited distances. A great circle represents the shortest distance between two points on the surface of the Earth, and can be visualized as shifting the equator until it coincides with the two points. The meridians of longitude are great circles which pass through the poles and are cut in half by the equator. A path traced along a great circle is merely a circular rotation around the center of the globe, using the radius of the globe. Although the shortest course for a ship between two ports is along a great circle, it is difficult to navigate because it requires a gradual, continuous change in compass direction. However, a practical course laid out by a navigator with one or two compass changes would very closely approximate a great circle route. This illustrates the close similarity of the constant bearing directionals to great circles over limited distances.

There is another rotational curve which is widely used in analyzing Plate Tectonics. It is based on the Fixed Point, or Euler's, Theorem which was developed by the mathematician Euler some 200 years ago. The theorem states that the most general displacement of a rigid body, turned about one of its points taken fixed, is equivalent to a rotation about a line through that point. As applied to the globe, the fixed point can be taken as the center of the Earth, and the rigid body as a specified area with a defined shape on the surface of the Earth. It is assumed that the area has shifted across the surface, and only the location of the specified area before and after the shift are known. The theorem considers only the two positions, and says nothing about the actual physical path which was taken to move from one position to the other. This is contained in the term "most general displacement," in which displacement refers only to the relationship between the two positions. The physical path followed from initial to final position could be straight, curved, spiral, or meandering and the displacement would still be the same. This can be visualized in the movement of a chess piece up, over, and down, compared to the displacement which is merely a line on the chessboard between the squares of the initial and final positions. "Most general displacement" on the Earth means that if any area moves from one place to another on the globe while maintaining its shape, the mathematical relationship between the beginning and end positions can always be represented by Euler's Theorem. This relationship can be visualized as a rotation of the area about some line through the center of the Earth that traces out a curved path across the surface between the initial and final positions of the specified area. A great circle curve is merely a special case of this theorem in which the rotation radius is equal to that of the Earth. Therefore, the constant bearing curve which approximates a great circle over limited distances would also approximate the most general displacement of the Fixed Point Theorem.

In a sense, almost any smooth line drawn on the surface of the Earth can be considered rotational because of the intrinsic roundness of the globe. This should be kept in mind when working with the constant direction straight lines on the flat Mercator Projection Maps which are used throughout the book. These straight lines are in reality curved lines on the true Earth which generally approximate both great circle curves and fixed point rotational curves along the shorter segments.

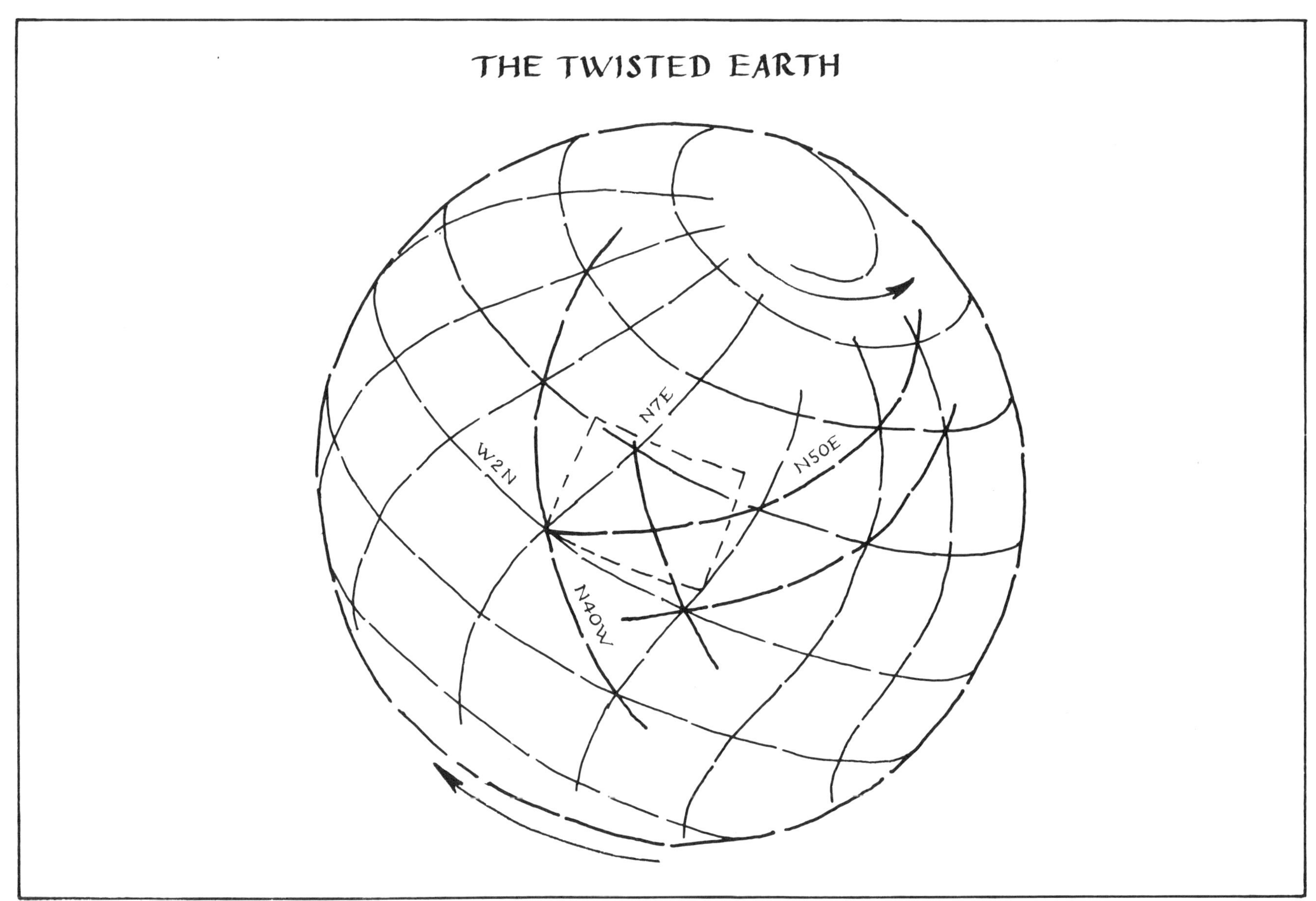
THE TWISTED EARTH
N7E
W2N
N50E
N40W

An analogy has been presented as an explanation for the four observed directionals which form a pattern across the surface of the Earth. The analogy considers a pure shear stress which twists the Earth around its axis through a 7.2 deg angle, creating a strain pattern in the crust. When projected on Mercator Projection, a pattern of unit squares deforms into a pattern of rhombus figures tilted 7.2 deg from the vertical and 1.8 deg from the horizontal, with maximum deformation along the diagonals. The similarities between the strain pattern analogy and observed directionals are many. Within the accuracy of observation, the orientation of the observed directionals and the strain pattern are the same. In addition, the degree of clarity between directionals matches the degree of deformation within the strain rhombus. The most clearly observed directionals are N40W and N50E which coincide with the diagonal lines of peak distortion in the strain rhombus. Less clear is the N7E directional which matches the less distorted uprights of the rhombus, and least clear is the W2N directional representing the least distorted top and bottom of the rhombus. The observed hierarchical nature of the directionals is also duplicated in the strain rhombus pattern. Lines of least distortion are the 7.2 deg lines off the vertical and 1.8 deg lines off the horizontal that cross the central diagonal intersection. They divide the strain rhombus into four similar figures, each of which can again be divided in the same way. Each time the operation takes place the sides of the rhombus figures are cut in half, setting up the sequence 1/2, 1/4, 1/8, and so on. Using the sequence, the vertical and horizontal tilt can be determined from the tangent of 7.2 deg equal to 1/8 and the tangent of 1.8 deg equal to 1/32. This gives strain rhombus diagonal angles of 40.5 deg and 49.5 deg which are very close to the observed N40W and N50E directionals. Using the hierarchical sequence to divide the full 360 deg width of the global Mercator Projection, the intervals 22.5 deg, 11.2 deg and 5.6 deg appear, matching the observed intervals found between directionals. If the analogy is tentatively accepted as explanation, it can be examined in somewhat more detail to develop a rational for variations in observed patterns.

The Earth's crust is made of porous rocky material which contains various fluids. Although the rocky material is fairly brittle, it has a certain amount of elasticity so that some deformation can occur before the rock breaks or fractures. It is this elasticity that allows the strain pattern to develop. The distortion or deformation in the crust increases or decreases as a directional is approached, until it reaches a high or low along a strain line. Along the pattern lines of peak strain the crust would preferentially tend to slip, slide, and tear apart, and along lines of least strain the crust would tend to be least disturbed. In both cases the lines would represent paths of high fluid mobility along which the fluids on and within the crust would tend to flow. On the surface are the streams, rivers, and lakes of fresh water which are supplied from rainfall, and below the surface are the shallow water aquifers which supply wells, and the deeper slow moving fluids of water, oil, gas, and molten rock. As fluid flows, it dissolves out minerals from the rock it contacts, eventually becoming saturated given sufficient time. The flowing fluid with its complex of dissolved minerals may pass through many different environments of changing temperature, pressure, and rock composition, along with exposure to other liquids and gasses. As environment changes, minerals in solution may be released to dropout and clog up the flow path, or additional minerals may be dissolved and the flow path enlarged. The action is complemented by other processes such as physical scouring to enlarge, or dumping of particles to constrict the path, and the movement of the ground during earthquakes may also open up or close fracture channels carrying fluids. The complex interaction of these processes on the fluid flow paths in the crust results in an intricate pattern of fluids moving through many different kinds of openings.

Along the preferential flow lines of the strain rhombus pattern it seems probable that the processes described above would be somewhat exaggerated, so that a flow pattern would carry within it a reflection of the strain pattern. The lines of extreme strain might tie to Earth features in two distinct ways. If the flow paths along the strain lines were enlarged, the features would tend to be aligned parallel to the directionals, and if the paths were restricted or blocked, the strain lines would act as a watershed with features directed away from the directionals. But nature is never so clearcut that it chooses only two distinct ways to develop a pattern. It seems likely that the preferential flow paths would open and close from time to time, forcing the flow to deviate and change direction as the path became clogged up. Although the flow might wander when first forced to change, it would eventually encounter another open preferential flow path in the near vicinity. The shift between two preferential flow paths would probably lead to curved tracks between directionals of different orientation, or offsets between those of the same orientation. However, the overall pattern or composite grain of the surface feature would be expected to reflect the strain pattern with its four directionals and hierarchical intervals.

In the next chapter the strain pattern will be tested by comparing it to the pattern of features within a small area. In addition to coastlines, the examples will also include fault systems, lava flows, and stream courses. Discussion will include both a description and general explanation of pattern relationships.

3 THE ISLANDS

DEVELOPING THE PATTERN

Examples in Chapter one were selected from the entire globe and were restricted to coastlines in order to simplify the presentation and reduce bias. It was found that the four directional trends in coastline examples were widely distributed across the surface of the Earth. In the last chapter a mechanical analogy was presented as an explanation, with a discussion indicating that a reflection of the directional rhombus pattern could possibly be found in flow features. This possibility was developed from the concept of preferential flow paths that coincide with lines of extreme deformation in the strain pattern. Along these directional trends the ease of flow could vary considerably as the path was opened up or clogged shut from place to place, so that rather abrupt changes in orientation could occur as the moving fluid shifted from one open directional to another, creating angles and offsets along the flow path. A test of this concept will require a wider variety of examples that are more directly related to flow, such as water and lava flow patterns. It might also be expected that fracture patterns that are directly related to crustal strain would also reflect the directional strain pattern. A stream course merely follows topographic lows, resulting in a simple line on a map. However, the characteristics of volcanic lava flow and fracturing add complexity to the picture.

Molten rock called magma flows upward from deep within the crust until it breaks out through a vent or rift on the surface as lava. Flow from a vent often creates a cone that can vary in size from a small hill to an entire mountain. Crater depressions can also form as material is carried away below the surface and the ground above collapses. It is not unusual that rift zones extend outward from a vent marking preferential flow paths below the surface along which magma moves, breaking out from place to place to form streams of molten lava. The flow of magma within the crust also tends to push up and swell the surface above as material is added from below, and eventually the surface rock may split apart into a pattern of fractures. Any strain network within the crust would tend to constrain the fracture pattern, which in turn would reflect the strain pattern. However, the correspondence between the strain and fracture patterns could be somewhat obscure due to the complexities of fracturing. A material in strain does not necessarily break along a line of peak strain, but rather tends to follow a zone of imperfection in the material. Fracture pattern examples will be discussed to determine just how they correspond to the strain pattern.

In view of the increased variety of surface features to be discussed, it is logical to initially narrow the field of study and concentrate on a relatively small area. Selection of the particular study area can be based on a logical set of criteria. A relatively small and isolated area is required so that it can be considered as a whole, somewhat removed from the rest of the world. A volcanic area is necessary, preferably with active volcanoes and recent lava flows that can be mapped more definitively than older, partially eroded flows. A long length of coastline to be used in tracing directionals would also be preferable. A group of islands would probably meet these criteria. The area requires good rainfall and fairly high elevations so that stream courses will stand out clearly and distinctly. It is also important that current and accurate information is available, which will probably require a well established and active research program. In summary, the study area might best be a small group of mountainous islands with active volcanoes, lots of rain, and with well documented information based on a high level of research effort.

The Hawaiian Islands meet all the above criteria. They lie in the Pacific Ocean some 2,000 miles from North America, the nearest large land mass. They are made up of a string of volcanoes that rise up from the sea floor to heights of nearly 14,000 feet above sea level. On the Island of Hawaii, the largest of the group, there are two active volcanoes that have erupted frequently over the last 200 years, discharging huge amounts of lava. Mauna Loa is the largest single mountain in the world, rising about 33,000 feet above the ocean floor, with its top 13,600 feet above sea level. In its 1950 eruption it pumped out about 200 million cubic yards of lava in a day and a half. Kilauea, with an elevation of 4,000 feet above sea level, is perched on the east flank of Mauna Loa and is known as the drive-in volcano. Over the last century it has erupted about every ten years on the average. The Hawaiian Volcano Observatory on the caldera rim of Kilauea has been actively monitoring volcano activity for more that 75 years. They routinely publish research reports and maps of the volcano activity that are available to the general public. The eight major islands in the chain have extensive coastlines and are in the path of the trade winds that drop up to several hundred inches of rain each year on the windward mountain slopes. Numerous streams flow down these slopes and discharge into the sea. In this chapter the Hawaiian Islands will be studied in relation to the directional strain pattern. The Pacific Basin setting will first be discussed, and then attention will be directed to geographical features on the islands.

3-1 Hawaiian Islands

From here on, all the illustrations will show a horizontal scale measure given in both degrees and distance. The degree scale is consistent with Mercator Projection and is correct along a horizontal over the entire mapped area. The degree scale intervals are taken from the hierarchical sequence shown in the last chapter, which was determined by dividing repeatedly by two, beginning with a full 360 deg circle. The distance measured on the scale is approximately true along a horizontal line that crosses the middle of each illustration. Due to the Mercator Projection characteristics, the distance scale can be considerably in error in the upper and lower portions of the figures. Each illustration also shows selected portions of a grid overlay made from the four directionals N7E, N40W, N50E, and W2N. The grid intervals between directionals are taken from the hierarchical sequence and are consistent with the degree scale shown on each figure.

Most of the maps in this chapter show coastlines and stream courses that were traced from "The Atlas of Hawaii," published by the University of Hawaii Press. Most of the fault patterns and historic lava flows were traced from "Volcanoes In The Sea," by Gordon McDonald and others, and then adjusted to fit the maps. Although the maps are not Mercator Projection, the correspondence is very close because the island group is near the equator and covers a relatively small area. As discussed before, distance distortion in Mercator Projection is greatest near the poles and least near the equator, and the Hawaiian Islands lie only about 20 deg north of the equator where distortion is minimal. It will also be remembered that in the conformal Mercator Projection small areas retain true shape, and the eight major Hawaiian Islands are spread over only 400 miles. On the extreme northwest is Kauai with its small neighbor Niihau just west. Moving across the Waho Shelf to the southeast is Oahu where most of the people of Hawaii live. Then comes the island group of Molokai, Lanai, Kahoolawe, and Maui, with the volcano Haleakala on east Maui rising to more than 10,000 feet above sea level. On the extreme southeast is Hawaii Island which has the same name as the entire group of islands. To avoid confusion, the island group will be called "The Hawaiian Islands," and the single island, "Hawaii Island." This island is also commonly called "The Big Island," because it contains almost two-thirds of all the area in the entire island group. Hawaii Island has several high mountains, with Mauna Kea and Mauna Loa reaching almost 14,000 feet and often snow capped in the winter.

Many of the island coastlines trend along directionals. The southeast coast of Hawaii Island in the lower right follows N50E with an offset at the center. On the west the trend is generally N7E, broken by two bulges at the center and north. The northern bulge resembles a nose which points along a general N40W trend towards the central island group to the northwest. This N40W general trend changes abruptly at Molokai, shifting west along the Molokai W2N trend across Penguin Bank. At Oahu the trend shifts back to N40W and can be extended to the northwest out along the Waho Shelf. The same directional shift which occurred at Molokai happens again north of Oahu where the N40W trend suddenly shifts to W2N, reaching for Kauai. The overall picture appears to be a series of stair step directional shifts alternating between N40W and W2N. In the upper left the small island of Niihau branches off from Kauai along N50E.

The subsea contours around the Hawaiian Islands represent ocean floor topography, and are shown as lines broken up by groups of dots. The 1,000 fathom contour is represented by breaks of one dot, the 2,000 fathom contour by breaks of two dots, and the 3,000 fathom contour by breaks containing groups of three dots. The coastlines of the islands are the heavy solid lines which trace the zero contour at sea level. The contours around the islands reinforce the overall trends discussed above. The N50E trend along the southeast coastline of Hawaii Island shows up clearly in the contours, and north of Molokai the W2N trend is evident. The contour around Waho Shelf northwest of Oahu follows N40W, and the N50E branch linking Niihau with Kauai is also more clearly defined by the undersea contours. To the north and south of the islands are several isolated, circular subsea contours, sometimes elongated. These represent small undersea mountains called seamounts which are scattered over the floor of the Pacific Ocean. Over recent years samples have been obtained from these seamounts which show them to be volcanic cones with flattened, eroded tops. South of the islands the N50E directional trend shows up clearly in the Bishop and Brigham Seamounts on the left and the Cross, Washington, and Perret group on the right, where the directional can be extended to trace the channel separating Hawaii Island from Maui. A N7E directional trend follows the McCall and Jaggar Mounts to the north where it separates Molokai and Lanai from Maui. The N50E trend is repeated with the Tuscaloosa Mounts north of the islands. In the next illustration the contours over a larger area of the Pacific will be shown and the directional trends discussed.

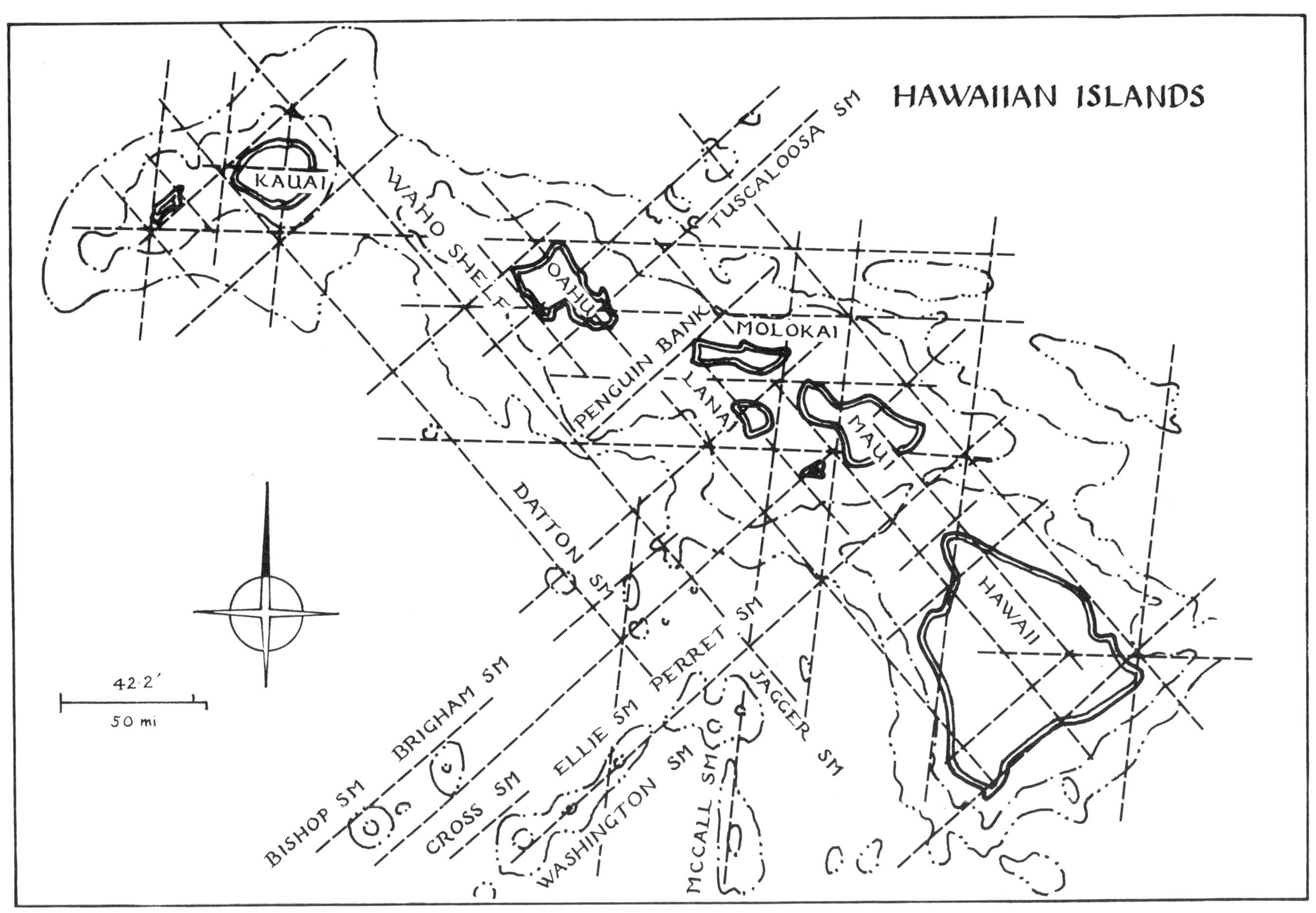

HAWAIIAN ISLANDS
KAUAI
OAHU
WAHO SHELF
PENGUIN BANK
MOLOKAI
LANAI
MAUI
HAWAII
TUSCALOOSA SM
DATTON SM
PERRET SM
JAGGER SM
BISHOP SM
BRIGHAM SM
CROSS SM
ELLIE SM
WASHINGTON SM
MCCALL SM
42·2'
50 mi

3-2 Pacific Ocean Floor

This shows a larger area of the north Pacific Ocean covering the entire Hawaiian chain from Hawaii Island in the southeast to Midway in the northwest, a distance of about 1,500 miles. The tracing is in Mercator Projection from a Department of Defense map and shows undersea contours as solid lines and island coastlines as double lines. The grid interval shown in the scale on the lower right is 2.81 deg, representing a distance of about 200 miles at the center of the figure. The Hawaiian Island chain is a string of huge undersea volcanoes that break the surface of the ocean in places to form islands. Most of the volcano mountain ridge is underwater, resting on the ocean floor about 15,000 feet below the surface. The largest mountain of the chain in terms of volume is Mauna Loa on Hawaii Island on the extreme southeast, which rises almost 14,000 feet above sea level. The enormous weight of this chain of mountains has pushed down the Earth's crust in the near vicinity, much like a wooden plank supported at the ends, which springs down when someone stands at the center. The saucerlike depression on the ocean floor south and east of the major islands is called the Hawaiian Deep, with the circular border of the saucer called the Hawaiian Arch.

As discussed on the last page, the eight major islands of the chain lie along a stair step trend alternating between N40W and W2N. On the southeast the trend starts with N40W along Hawaii Island and Maui, changes to W2N at Molokai, and then back to N40W at Oahu. Above Kauai the general trend again changes to W2N and continues on to Nihoa and Necker Islands, where a branch strikes off N50E to the southwest along Necker Ridge, much like the N50E Niihau branch that strikes off from Kauai. Just west of Necker Island the trend again changes to N40W and continues for about 200 miles, before changing back to W2N which it follows for about 300 miles, and then again shifting back to N40W along Midway and Kure Islands. These leeward

Islands, which run for about 1,200 miles along the directional stair steps from Kauai to Midway, are relatively small, highly eroded remains of volcanoes with many associated reefs or banks lying at or slightly below sea level. In the northwest, Pearl and Hermes Reefs and Midway and Kure Islands are atolls with a circular enclosing coral reef.

The undersea contours in the lower left of the figure also show evidence of the directional pattern. The N50E Necker Ridge trend is like an arrow pointing southwest along the undersea Pacific Mountain Ridge which continues on to the Cape Johnston Tablemount where there is an abrupt shift to N40W. East of Necker Ridge across an interval of 2.81 deg, the N50E trend also appears at Horizon Tablemount. On the southwest end of Horizon Tablemount the trend takes a 90 degree angle and strikes off towards the southeast along N40W, following the Karin and Sculpin Seamounts. The overall pattern that emerges appears to be a complex directional network of ocean features with many branches and offsets created by many abrupt changes in directional trends. The predominate directionals are the N40W and N50E diagonals which intersect at right angles. This is especially evident where the N50E Necker Ridge trend forms a spine out of which branch the two N40W directional trends of Karin and Sculpin Seamounts to the southeast and Cape Johnston Tablemount to the northwest. Although the N7E directional is not widely distributed, it shows up clearly just northwest of Necker Ridge where a closed contour in the Hawaiian chain is bordered by a N7E directional pair separated by a 1.41 deg hierarchical interval. Other N7E directionals also appear as dividers along the Hawaiian Island chain. As discussed above, the W2N directional shows up repeatedly in the Hawaiian Island stair step trend.

However, there are two areas of the figure where the contour patterns do not seem to reflect any of the four directionals. These areas occur at the top and bottom of the page on the right half of the figure. To

the north are the Musicians Seamounts, bordered on the west by a N40W directional that can be extended southeast to form one side of a square which boxes in Hawaii Island. East of the Musicians Seamounts is a group of contours that trend upward toward the upper right corner at an angle of about east 10 deg north. In the southeast corner of the figure is another group of contours which extend to the right hand edge, and have about the same orientation as the contour group in the north. This southern contour group is bordered on the west by a pair of N50E directionals that can be extended to the northeast to form two sides of the square enclosing Hawaii Island.

These two contour groups on the upper right and lower left of the page represent a portion of two long fracture zones formed of a large number of closely spaced shorter fractures. These two zones are part of a huge fracture system extending west from the Americas far out into the Pacific, and from Alaska south across the equator. The fracture system and its correspondence with the directional pattern will be described on the next page.

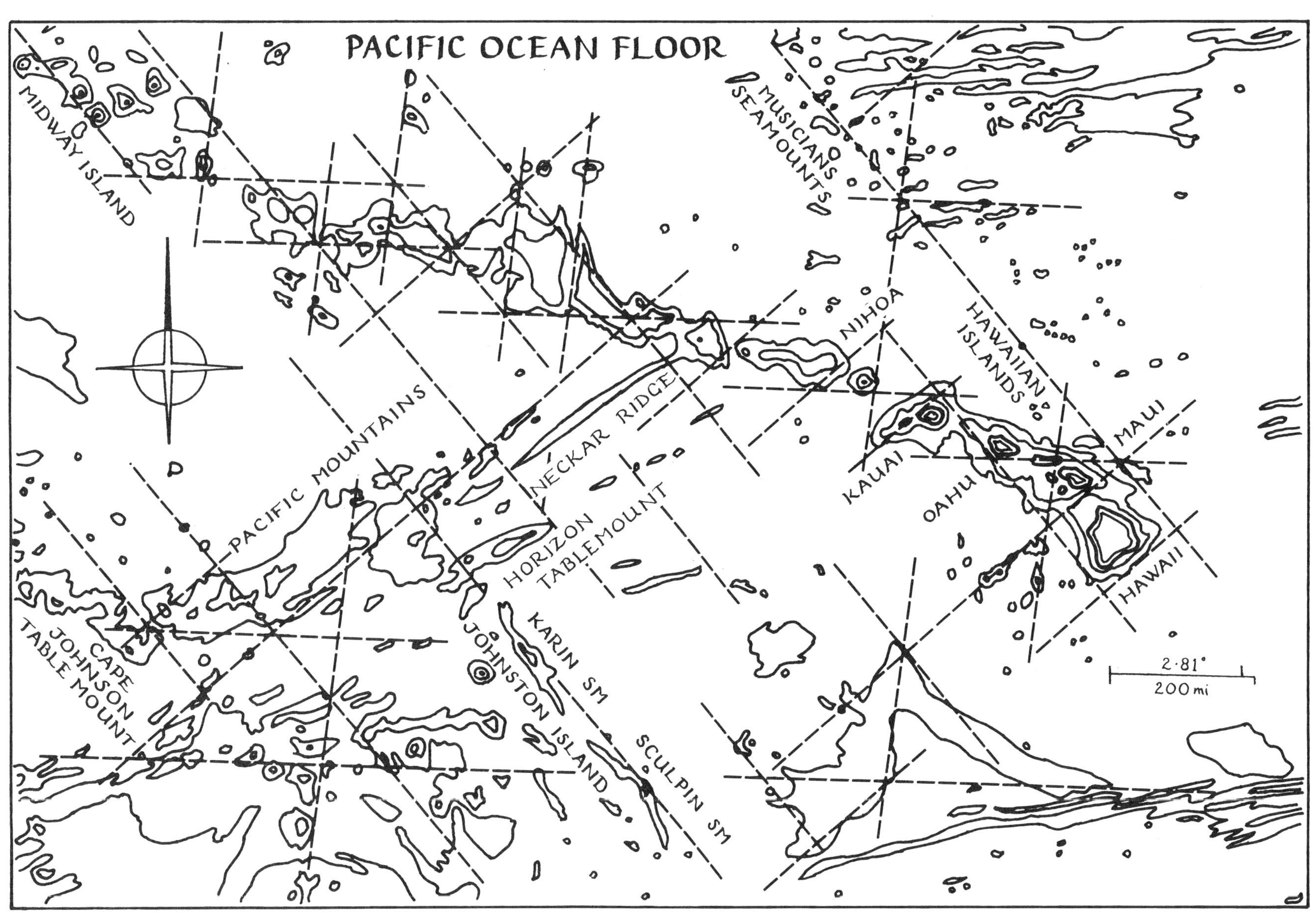
PACIFIC OCEAN FLOOR
MIDWAY ISLAND
MUSICIANS SEAMOUNTS
NIHOA
HAWAIIAN ISLANDS
PACIFIC MOUNTAINS
NECKAR RIDGE
HORIZON TABLEMOUNT
KAUAI
OAHU
MAUI
HAWAII
CAPE JOHNSON TABLE MOUNT
JOHNSTON ISLAND
KARIN SM
SCULPIN SM
2·81°
200 mi

3-3 Pacific Fracture Zones

This shows a large area of the Pacific from the equator in the south up to about 40 deg north latitude, and over 3,000 miles from east to west. In the upper right corner is the California coast of North America, and just left of center are the Hawaiian Islands. The five sets of lines which run westward across the Pacific from North America are fracture zones. Although the zones are shown by a single line, they are actually a closely packed group of shorter fractures up to 120 miles in width which are strung out along the fracture zone. The sea floor on either side of these zones has been shifted both laterally and vertically. The lateral shift along the zones is as much as several hundred miles, and the vertical displacement, which shows up as a difference in depth across the zone, is several thousand feet in some locations. Where there is a significant difference in depth across a zone the deeper or downthrown side is shown as toothed. The Clipperton Fracture Zone at the bottom of the figure is smooth over most of its length, indicating that the displacement is mostly lateral with one side sliding past the other, and there is little difference in depth across the zone.

There are a total of 13 fracture zones in the full set extending over 4,000 miles north to south, with individual zones running for up to 2,000 miles east to west. There is no other place in the world where a set of fracture zones of this size and extent exist. Starting from the north the five fracture zones shown are the Mendocino, Murray, Molokai, Clarion, and Clipperton. They can be visualized as fences which divide the eastern Pacific up into strips of about the same width when shown on Mercator Projection. Although there are major differences in the sea floor characteristics between strips, each individual strip seems to have relatively consistent properties. The strip north of the Mendocino Zone and the strip between the Murray and Molokai Zones have much the same character, with many volcanic seamounts scattered over a relatively thin crust. Between these two strips lies the one bordered by the Mendocino and Murray Zones with quite different properties. On the average, the sea floor in this strip is several thousand feet deeper than its neighbors, with very few seamounts and a thicker crust.

A directional pattern network with an interval of 5.62 deg has been drawn over the five fracture zones shown on the page. On the east, the ends of the top three zones are bounded by a N7E directional. Further south other N7E directionals, offset across hierarchical intervals, also bound the zone ends. On the west the pattern is similar, with the top four zones bounded by a N40W directional, and with additional offset directionals bounding the zone ends to the south. On both east and west, the zone ends approximately coincide with a group of W2N directionals separated by equal hierarchical intervals equivalent to an 11.2 deg horizontal interval. In further discussions the words equivalent and horizontal will be dropped with the understanding that the qualification continues to apply. Each fracture zone seems to angle across two adjacent W2N directionals, starting in the east at the upper W2N and dipping south and west to end on the W2N just below. All the fracture zones are somewhat broken up as they wander across the page, with the Molokai Zone showing the most breaks and the Clipperton the least. Many of these breaks coincide with various directionals, showing a correspondence with the hierarchical directional pattern which seems to serve as a framework on which fractures initiate and end.

It will be recalled from the introductory discussion of this chapter that when a material breaks under stress, the fracture trend generally follows imperfections in the material rather than the strain lines. This characteristic would also probably hold true for breaks in the crust of the Earth, so that fracture orientation could deviate considerably from the orientations of the strain pattern. The correspondence between a fracture system and the underlying strain pattern is of a different sort than shown before. The lines in a strain pattern resulting from a stress represent zones of extreme strain along which the material shows either peak or least deformation. When the strain becomes great enough for the material to break, it seems reasonable that the fracture would generally originate along a strain line where the deformation is at a high or low extreme. At intersection points, or nodes, where strain lines cross, there would seem to be an even greater probability of initiating a fracture. The correspondence between a strain pattern and a fracture system might be described as a constraint relationship, where the ends of a fracture are constrained by the lines in the strain pattern.

The constraint effect is readily apparent on the figure, with fracture ends constrained by N7E on the east, and by N40W on the west, and with both ends constrained within the W2N directional set. Other constraining effects appear at offsets and breaks along the zones. Additional fracture systems will be examined to further test these concepts.

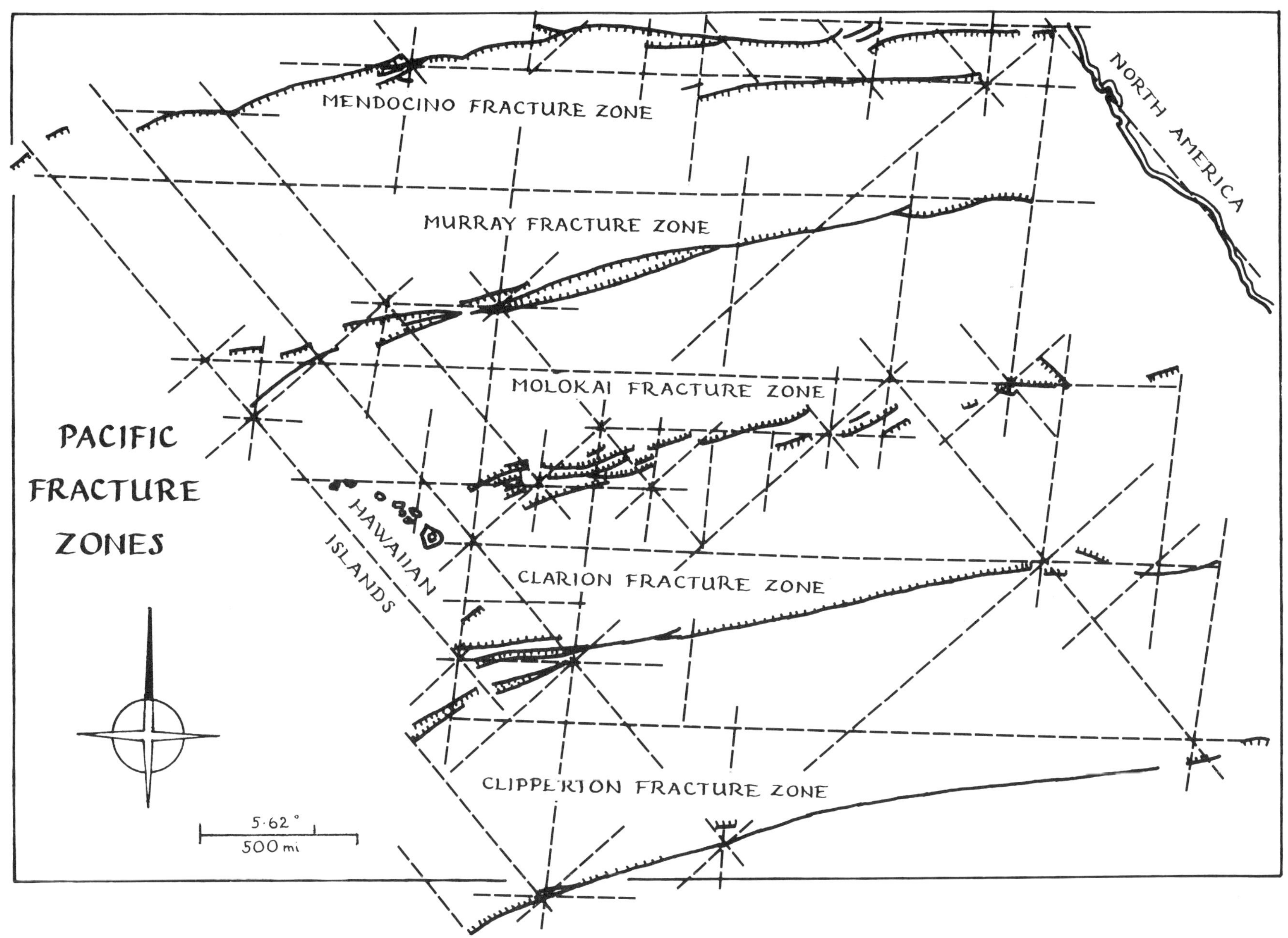
MENDOCINO FRACTURE ZONE
MURRAY FRACTURE ZONE
MOLOKAI FRACTURE ZONE
CLARION FRACTURE ZONE
CLIPPERTON FRACTURE ZONE
NORTH AMERICA
PACIFIC FRACTURE ZONES
HAWAIIAN ISLANDS
5·62°
500 mi

3-4 Pacific Trenches

This shows the configuration of the deep ocean trenches of the Pacific. The area covered extends about halfway around the world from Asia on the west to the Americas on the east, and from Alaska in the north to southern South America. The scale at the bottom of the page shows that 22.5 deg is equivalent to about 1,500 miles at the center of the page. Because of the large area covered on the Mercator Projection, the distance scale is considerably distorted in both the upper and lower portions of the figure. The deep ocean trenches shown were traced from "World Ocean Floor Panorama" by Bruce C. Heezen and Marie Thorp, and rescaled to fit the Defense Department Mercator Projection map. All coastlines except those of the Hawaiian Islands were eliminated so that the Pacific Trenches can be examined in isolation.

The deep ocean trenches are the largest linear features on the surface of the Earth. The trenches resemble a deep slash cut one to two miles into the ocean floor, with the bottom of the slash somewhat flattened due to sediment filling. The average width of the top of the trench on the ocean floor is about 50 miles, and on the landward side of each trench is either a continental coastline or island arc. On the lower right of the figure is the Peru-Chile Trench off the coast of South America which runs for almost 3,000 miles without significant interruption. Just inland, the Andes Mountain range rises directly out of the trench to form the west coast of South America. About 100 miles further inland within the Andes Mountains lies a string of volcanoes, many of which are active. Northwest of South America is the Middle American Trench which also borders a continental coastline with a string of volcanoes just inland. Along the west coastline of North America the circum-Pacific trench belt is broken and replaced by the San Andreas Fault which runs northwest towards Alaska where the Aleutian Trench lies just south of the Aleutian Basin and Island chain. This configuration of a deep ocean trench bordering an island chain with a basin behind it is typical of the Western Pacific. Southwest of the Aleutians is the Kuril Trench bordering the Kuril Island chain and Okhotsk Basin, and just below is the Japan-Bonin Trench with the Japanese Islands and Japan Basin to the west. South of Japan are the Mariana and Yap-Palau Trenches enclosing the West Philippine Basin, and separated from the South China Basin and Asian mainland by the Ryukyu and Philippine Trenches. Further southwest is the Java Trench bordering the Indonesian Island chain, and to the southeast are the Solomon, New Britain, New Hebrides, Tonga, and Kermadec Trenches, each in line with its own island chain.

This circum-Pacific belt of trenches, associated basins, and volcanic island chains or continental coastlines is coincident with the Pacific "Ring of Fire" where much of the world's earthquake and volcanic activity occurs. About 85 percent of all the earthquake energy released in the world comes from the circum-Pacific belt. These earthquakes tend to cluster along the ocean trenches, and deeper below the surface under the adjacent island chains or continental borders. This concentration of earthquake and volcanic activity indicates an enormous disruption in the Earth's crust as it deforms, fractures, and dislocates below the trenches. Each trench is essentially a fracture zone, much like those discussed on the previous page, and the entire complex of circum-Pacific trenches can be considered a fracture zone system which stretches across the entire Pacific. The correspondence of this huge fracture system to the directional strain pattern is much the same as discussed on the last page.

Several of the trenches are oriented along directionals. On the east, the Peru-Chile Trench trends north along N7E, then shifts abruptly to N40W, and then back to N7E at its northern end. Further north the Middle America Trench branches off from a N40W directional and gently curves up across a 11.2 deg interval, then straightens out to follow another N40W directional. Across the Pacific to the west, the Kuril and Ryukyu Trenches follow N50E, and the Mariana Trench forms a rounded right angle with its two sides trending N40W and N50E. However, the principal pattern characteristic is the tendency for the trenches to begin, end, and break at directionals, so that the directional grid acts as a network constraining the pattern of the trenches. Several of the constraining directionals may be followed across the entire Pacific. The two abrupt orientation changes in the Peru-Chile Trench are located on two W2N directionals that can be extended west to intersect breaks in the Solomon Trench complex, and on to a break in the Java Trench. The east end of the Aleutian Trench lies against a N50E directional which extends southwest to constrain the Mariana and Yap-Palau Trenches. Another N50E directional can be traced from the north tip of the Middle America Trench to the break in the Tonga and Kermadec Trenches. The fit of the west Pacific trenches across hierarchical intervals is also evident in the rectangular nature of the trench pattern constrained by N40W and N50E directionals. A much smaller area will be studied next to further test the constraining character of directional grid.

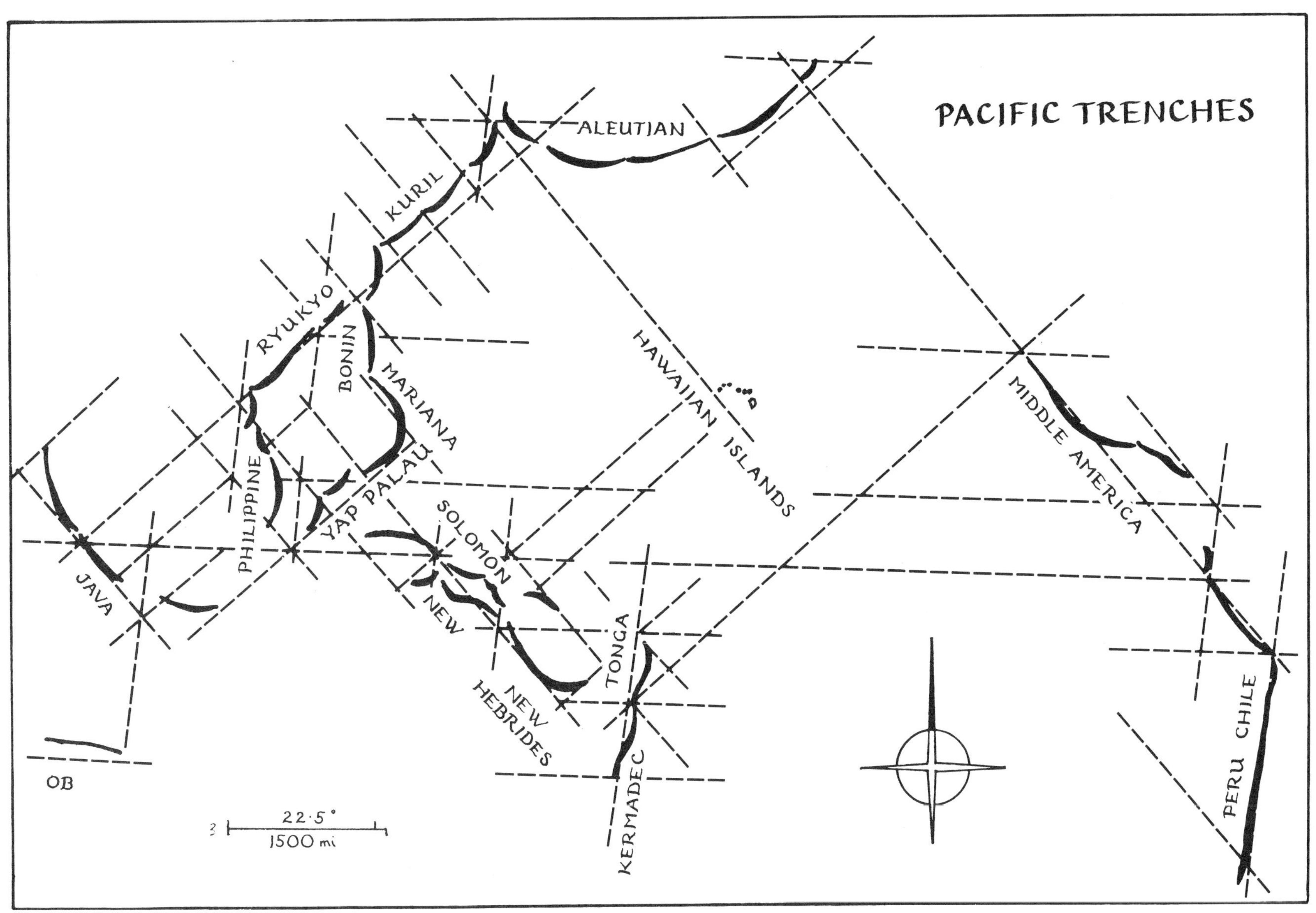

PACIFIC TRENCHES
ALEUTIAN
KURIL
RYUKYO
BONIN
MARIANA
PHILIPPINE
YAP PALAU
SOLOMON
NEW
NEW HEBRIDES
TONGA
KERMADEC
JAVA
OB
HAWAIIAN ISLANDS
MIDDLE AMERICA
PERU CHILE
22·5°
1500 mi

3-5 Kilauea Fractures

This shows the major fracture patterns of Kilauea Volcano on the Island of Hawaii. Kilauea Volcano is a gently sloping shield volcano in the shape of an overturned saucer covering the eastern portion of Hawaii Island. It rises up about 20,000 feet from the sea floor with the top 4,000 feet above sea level. At its summit is an oval depression about three miles long and two miles wide called a caldera, and within the caldera is a crater called Halemaumau, which for many years held a liquid lava lake. Halemaumau is the home of the beautiful, but easily angered, volcano goddess Madame Pele who releases streams of lava when she cuts into the ground with her wooden hoe, and cracks the earth when she stamps her foot. She is a lady who should be treated with respect and is not to be trifled with.

Although Kilauea sits on the southeast flank of the huge Mauna Loa Volcano, there seems to be no connection between the volcanic activity of the two. Two rift zones branch out from the Kilauea summit caldera. These zones are made up of many small fractures with the same general orientation which are clustered together forming preferential flow paths for underground magma moving outward from below the caldera. The Southwest Rift Zone begins just left of the caldera and angles down towards the southwest, gradually curving in to the south along the Great Crack as it approaches the ocean. Close to the caldera, the rift zone with its many fractures is bounded by two N50E directionals separated by a 2.64 min hierarchical interval. As the Southwest Rift Zone approaches the sea it seems to shift gradually towards N7E, although lava flows along the rift, which will be shown later, indicate that the directional shift may be more abrupt than is apparent from the fracture pattern. The Southwest Rift is separated from the triangular Koae Fault System by a 2.64 min interval oriented N50E which shows little fracturing. On the east, the Koae system is bounded by a N40W directional which traces the initial path

of the East Rift Zone as it leaves the caldera. This Rift Zone runs for only a short distance along N50E before it abruptly shifts eastward on its way to the sea. The directional configuration will be examined further in another figure showing recent lava flows along its path. These two rift zones, which branch off from either side of the caldera and provide preferential flow paths for magma, constitute one type of fracture system.

The Kilauea Volcano grows in two ways. Magma from the mantle forces its way upward from deep in the Earth to collect just below the caldera. Some of this magma erupts out on the surface as lava, which slowly piles up in layers to raise the height of the mountain. Magma also moves underground through the rift zone fractures, forcing them open as it pushes its way along. Much of this magma stays below the ground and hardens into rock within the widened fractures, adding area laterally to the volcano. Kilauea is jammed up against Mauna Loa on the north so that the block of land south of the rift zones tends to be pushed south as the zones widen, with chunks of the leading edge sloughing off and slipping down the south flank to form stair step cliffs, called pali in Hawaii. The Koae and Hilina Fault systems show the patterns of these stair step fractures formed as the south block gradually moves oceanward. As noted before, the Koae Fault system is enclosed by a directional triangle with W2N on the bottom, and the upper sides formed with N40W and N50E directionals which intersect in a right angle just southeast of the caldera. The Hilina Fault system is bounded on the north by a W2N directional and on the west by a N50E directional, which is the furtherest south of a set of four separated by 2.64 min. Starting from the north and working south, the first N50E defines the upper border of the Southwest Rift Zone and the second its lower border. The third follows the upper left border of the Koae Fault System, and the fourth the west border of the Hilina Fault System. Another set of four directionals trending N40W and separated by 2.64 min intervals also crosses this area.

The first in the northeast issues from the caldera to follow the beginning of the East Rift Zone, and the other three to the southwest are defined by breaks and offsets in the individual fractures of the various fault systems.

Although the area considered is much smaller, the correspondence between the directional grid and fracture zones follows the same characteristics here as in the previous two figures. The scale on the previous figure gives a unit distance of 1,500 miles which is about 300 times greater than the unit five miles given on this figure. The individual fractures are rarely oriented along any one of the four directionals of the strain pattern, but rather follow probable imperfections or anomalies in the structure or composition of the material. As before, the fractures tend to begin, end, and branch on a directional, so that the constraining directional pattern acts as a framework supporting the fractures. These characteristics are often observed as directional borders enclosing fracture systems. On the next page a fracture system in an even smaller area will be examined.

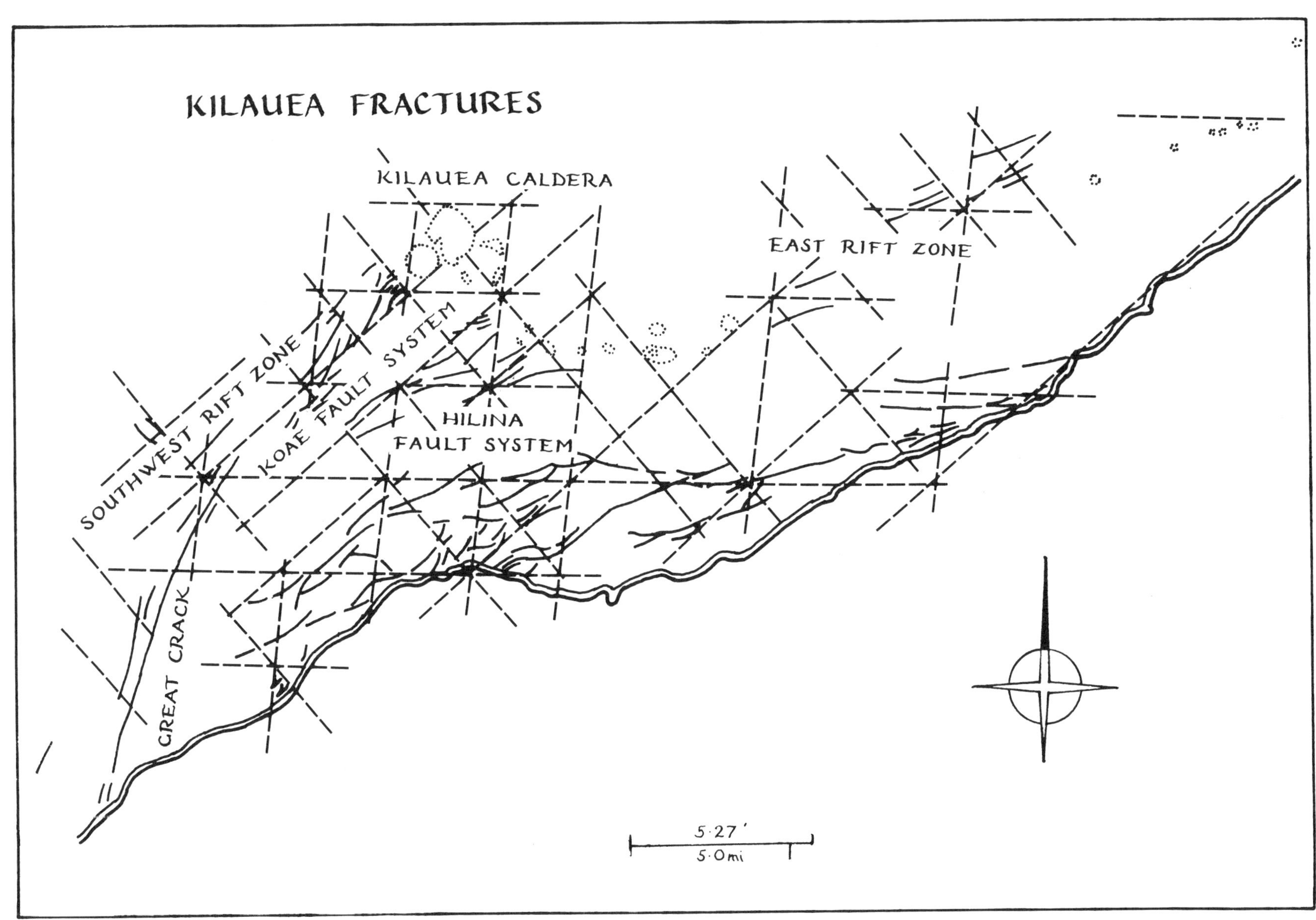
KILAUEA FRACTURES
KILAUEA CALDERA
EAST RIFT ZONE
SOUTHWEST RIFT ZONE
KOAE FAULT SYSTEM
HILINA
FAULT SYSTEM
GREAT CRACK
5·27'
5·0 mi

3-6 *Nihoa Dikes*

This shows Nihoa Island which lies about 170 miles northwest of Kauai, and is the farthest south of the Leeward Hawaiian Islands. The scale shows that 1,000 feet approximates 9.89 sec, which compares to about five miles on the last figure. Although Nihoa is less than a mile long, there are remains of house sites and garden terraces indicating that people did live on the island at one time. It must have been a lonely and hard life on an island that could be walked across in 15 minutes, and where the nearest neighbors were several days away by boat. Nihoa is made up of thin layers of lava which are the leftovers of a shield volcano that has largely eroded away. The solid lines on the figure represent dikes, which are fractures filled with magma that flowed into the cracks and became solid.

This process of filling and solidifying within fractures can vary, depending on fracture depth. If a fracture is near the surface, the process may be a simple flow of magma into the cracked rock, with the magma gradually slowing down and finally hardening as it cools. However, for fractures at depths the high pressures and temperatures can modify this simple process considerably. Pressure usually increases about one psi for each two feet of depth, so that at one mile below the surface a pressure of about 2,640 psi might be expected. The increase of temperature with depth varies widely over the Earth but can be extremely high in the vicinity of molten rock. The character of liquid flow within rock depends on the viscosity or thickness of the flowing liquid and the size and type of the openings in the rock. Molasses flows slower than water because it is more viscous, but any liquid will flow faster through a large opening than through a small hole. Most rock is not completely solid, but has a complex of holes made up of pores or fractures. If these holes are relatively large and connected, fluid may easily flow through the rock, but if the holes are small or the fractures closed and each opening is separate from the other, flow

would be impossible or very difficult. The network of openings in an underground rock is always filled with something, so that any fluid that leaves the network is always replaced by some other fluid that enters the network. In addition, if the rock should fracture or break so that more space is created in the network of holes, more fluid from outside must enter the rock to fill up the increased space.

Fracturing within the Earth's crust can happen almost instantaneously. Material under stress will gradually deform until it suddenly breaks and the strain or deformation is relieved. As the rock is torn apart, space is usually created along the ragged and offset edges of the fracture. Although a new, partially open fracture will tend to close rather quickly as the rock readjusts under pressure, there is a finite time that the newly created space with its extremely low pressure will stay open. During this short finite time, any high pressure fluid in the near vicinity will violently expand and force itself into the new opening at a very high flow rate. This explosive blast of fluid into the new fracture from neighboring rock would generally happen with enormous variations in pressure and probably temperature, depending on the expansion characteristics of the fluid which could be molten rock containing a complicated mix of materials. Extreme variations in pressure and temperature associated with the complex magma mix could release gases, selectively crystalize minerals, or dissolve adjacent rock, changing completely the character of the residue remaining in the newly opened fracture. Depending on the circumstances, solid deposits such as crystals could plug up the fracture or hold it open to be filled with magma that would eventually cool and solidify forming a dike. Over time, the material above would gradually erode away until the filled fracture appeared at the surface. The dikes on Nihoa were probably formed much as described above, and represent the fossilized image of fractures which may once have existed at a great depth.

Directional trends in both coastlines and the dike fracture system are evident. On the southwest the coastline forms a rectangular peninsula leg which is bordered by two N40W directionals separated by a 4.94 sec interval. In the center of the southern coast is an inverted "V" formed by intersecting N40W and N50E directionals. The southeast coast trends N50E, and the north coast follows W2N with the center inset to the south. Although the general trend of the dikes is N50E, most of the individual fractures diverge somewhat from the N50E directionals. As in other fracture systems studied, the most obvious characteristic is the tendency for fractures to begin, end, and change direction on various directionals. It is also observed that many fractures end on directionals that follow coastlines, which could indicate the tendency of erosional processes to be constrained by the directional pattern.

In the next few figures, recent lava flows on the Hawaiian Islands will be examined in conjunction with the directional pattern. These flows tend to follow the natural grain of the land along preferential flow paths as they move towards the sea. Before discussing the correspondence between these flows and the directionals, an overview of Hawaii Island will be presented.

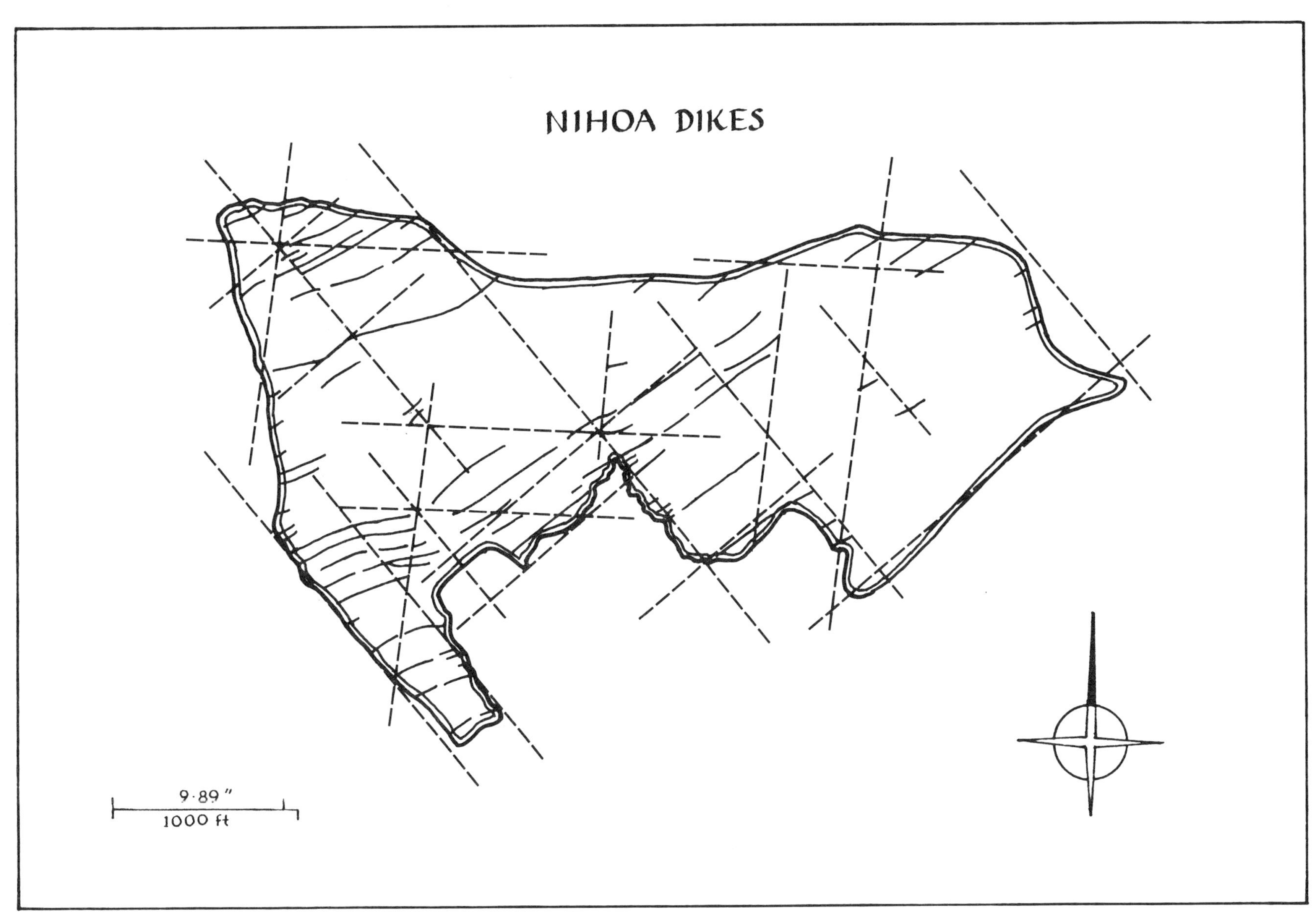

NIHOA DIKES
9·89"
1000 ft

3-7 Hawaii Lava Flows

Hawaii Island is shown here with historic lava flows and linear trends. The figure is a composite, using a number of sources that include, "Atlas of Hawaii," "Volcanoes in the Sea," and a satellite photo. The island is formed of five volcanoes, with Mauna Loa in the center flanked by Kilauea on the east and Hualalai on the west, all of which have erupted in historical times. To the north lie the two dormant volcanoes Mauna Kea, located just above Mauna Loa, and the Kohala Mountains that form the extreme northwest nose of the island. In the northeast is the town of Hilo, a small agricultural city of about 40,000 people that overlooks beautiful Hilo Bay and has as a backdrop the 13,800 foot Mauna Kea, sometimes snowcapped in the winter. The wet Hamakua coast stretches northwest of Hilo to the Kohala Mountains, with the coast road running high above the ocean along the north flank of Mauna Kea, passing through the sugar cane fields and across the deep-cut valleys. The streams on this windward coast are fed by the moisture loaded tradewinds that blow in from the northeast to flood the slopes with heavy rainfall that finds its way to the ocean down stair step waterfalls through the rainforest valleys.

As the Hamakua coast road approaches the Kohala Mountains, it cuts southwest and runs inland, climbing the lower Mauna Kea slopes to reach the small rural town of Waimea, located on the open pastures of the cattle country. Much of the land here is part of Parker Ranch that covers the northwest flank of Mauna Kea on the left and the misty green Kohala Highlands on the right. The road continues southwest, dropping down to the sunny dry South Kohala and North Kona coasts with their black lava fields, and then curving around Hualalai towards the resort town of Kailua, Kona. The road leaves Kailua to run south high above the ocean on the west flank of Mauna Loa, and then curves east around the southern end of the island, passing over numerous lava flows from Mauna Loa's southwest

rift. On the right of the road is South Point, the southernmost area in the United States where stair step cliffs called pali mark the fracture zones. This is a lovely, lonely land, with its dark lava flows, green grasslands, and dry scrub areas.

Between South Point and Kilauea is a green pocket, filled with cane fields, macademia nut orchards, and dairy farms, lying between the black sand beach of Punaluu on the ocean side and a number of grassy hills perched on the flank of Mauna Loa. These hills are the eroded remains of the ancient volcano Ninole, most of which has not been covered with Mauna Loa lavas. The major town in the area is Pahala, which has the dubious honor of once recording a temperature of 100 degrees Fahrenheit, the highest ever reached in the Hawaiian Islands. The road continues in a straight line, gradually climbing the slopes of Kilauea to the 4,000 foot summit within the Hawaiian Volcanoes National Park. At the summit the fascinating Crater Rim Drive branches off from the main road to circle the caldera, and at one point, dips down into the caldera to pass the Halemaumau firepit crater. Another branch called the Chain of Craters Road winds down the east flank of Kilauea towards the sea, cutting across the huge expanse of lava from the 1969 to 1974 flow that flooded over the Hilina Pali on its way to the sea. Most of this flow came from a vent now marked by the 400 foot Mauna Ulu cone built during the eruption.

At the coast the Chain of Craters Road joins another highway running northeast along the Puna coast. Although at this writing the road is blocked by lava and must be approached from a different direction, on the other side of the lava it continues on past the Kalapana and Kaimu black sand beaches, and then to the non-existent town of Kapoho that was buried in a 1960 lava flow. The sea along this windward coast road is almost always rough and rolls in to continually pound the high lava cliffs, eating away the rock to create graceful sea arches and isolated rocky stacks just offshore. In places, the road breaks

back from the coast to travel through tunnels of mango and pandanus trees, passing a number of parks for picnicking, or perhaps just quiet contemplation. From Kapoho the road turns northwest past anthurium nurseries and papaya farms, then through the country towns of Pahoa and Keaau and back to Hilo, completing the island circle.

Much of the coastline and many linear features trend along directionals. Hilo Bay in the northeast is bordered by W2N and N7E directionals, and to the north west the Kohala Mountains lie within two N40W directionals separated by a 21.1 min hierarchical interval. Below Kohala the coastline shifts abruptly to follow a N50E directional that crosses two N40W directionals defining two lava flows. Further down, the coast trends N7E, then curves around the south tip to stretch out towards the northeast along N50E, broken by an inset near the center. To the west are the Mauna Loa rift zones which are marked by lava flows. They also follow N50E, shifting abruptly to W2N towards Hilo on the north, and splitting to W2N and N7E on the south. Just above the center of the figure, a portion of the Mauna Loa lavas branch off to follow N40W towards the northwest along the contact between Mauna Loa and Mauna Kea. In the east the Kilauea lavas appear constrained by N40W and N50E directionals. This area will be examined in more detail in the next figure.

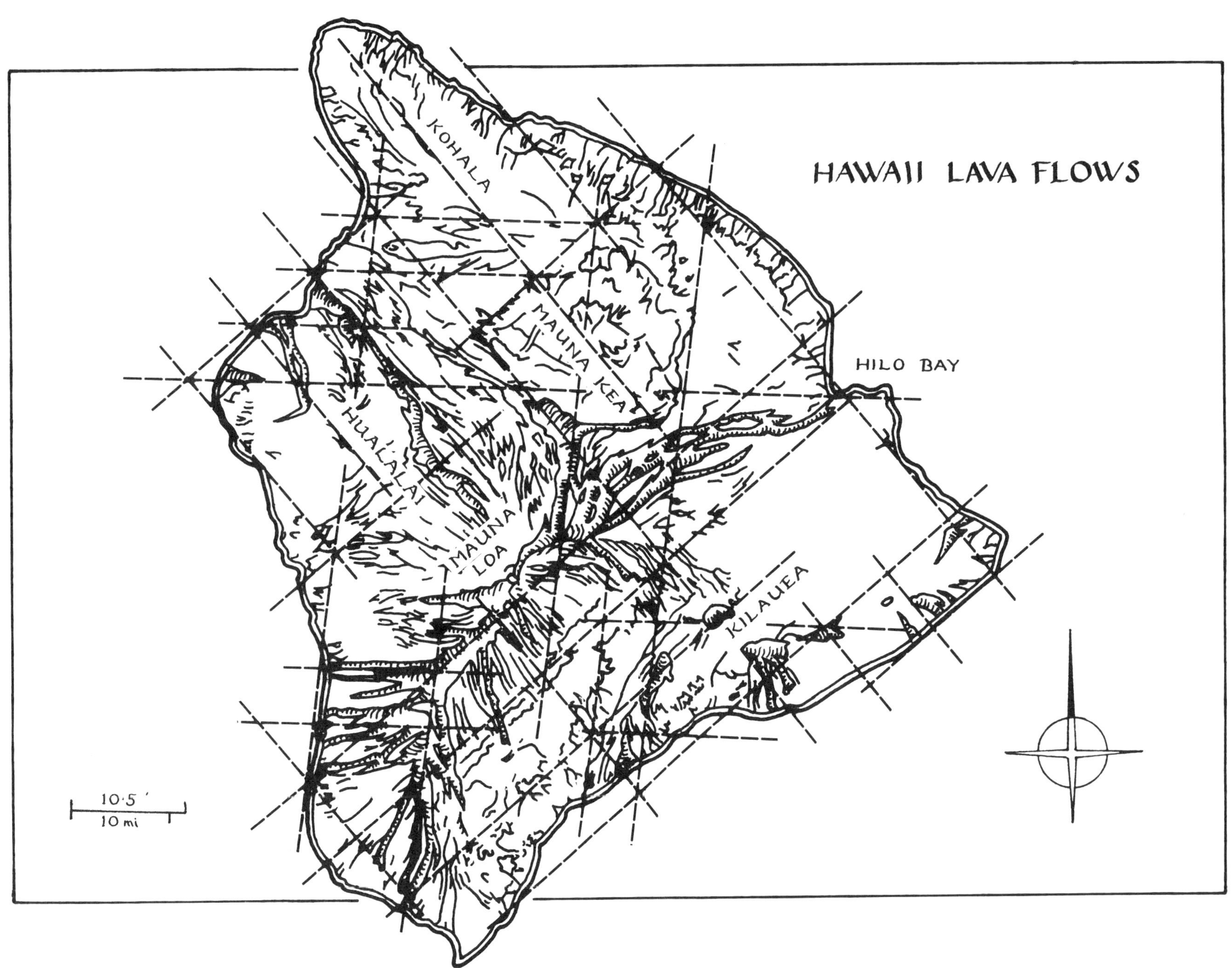
HAWAII LAVA FLOWS
KOHALA
MAUNA KEA
HUALALAI
MAUNA LOA
KILAUEA
HILO BAY
10.5'
10 mi

3-8 Kilauea Lava Flows

Historic lava flows from Kilauea Volcano are shown in this figure. Left of center is Kilauea Caldera with the Southwest Rift Zone stretching away from the caldera along N50E. Within this rift zone is the Great Crack that runs for almost 15 miles, gradually bending towards the south as it nears the coast. The lava flow in the extreme lower left oozed out of this crack in 1823 and was reported to have moved so fast on its way to the ocean that several people were caught in the flow. Further up the rift zone is the prehistoric Keamoku lava flow from Mauna Loa, marking the contact between the two mountains. Although this flow is not shown on the figure, it also follows N50E. Just southwest of the caldera are shown two lava flows that erupted from the Southwest Rift in 1920 and the 1970's. the two flows are separated by a 2.64 min hierarchical interval and show much the same configuration, starting from the caldera and running N50E for about 5 miles through an offset near the center, then abruptly shifting direction to follow N7E on the downstream end.

The two flows in the caldera are from short eruptions in 1971 and 1974 associated with the 1969 to 1974 Mauna Ulu activity that produced the large lava flows beginning just southeast of the caldera and continuing on to the sea. The activity began February 1969 on a fairly small scale with lava covering a portion of the Chain of Craters Road. The eruptions resumed on a larger scale in May and continued to July 1974, sending large streams of lava over Hilina Pali and into the sea several times, while destroying most of the Chain of Craters Road that ran from the caldera to the coast. The new Chain of Craters Road that was opened in 1979 runs for several miles across these relatively fresh lava flows. At the main vent near the caldera where most of the lava poured out, the Mauna Ulu shield cone formed and eventually reached a height of 400 feet. The west tongue of this flow follows N7E on its way to the sea, while the east tongue begins along N40W and then curves down towards N7E as it approaches the ocean. A smaller tongue of lava just northeast of the source branches off along N50E for a short distance, and just beyond it to the east is a 1965 lava flow with its base lying on a W2N directional. These flows mark the East Rift Zone that approximately parallels the shoreline on its way from the caldera to the sea. Just southeast of the caldera the rift zone follows a "V" centered on the Mauna Ulu cone, then strikes east along the W2N base of the 1965 flow mentioned above. When it reaches the Puna area, the zone shifts to the northeast along N50E, and then through a N7E offset before it reaches the sea.

At about the center of the Puna coast are several lava flows that strike N40W from the East Rift and continue to the ocean. The lower flow of this group occurred in 1750, and the other named flows above, along with the Kii flow to the northeast, erupted in 1955. This was the first volcano activity in the area for 115 years. The last flow prior to 1955, shown northwest of the Kii flow, had occurred in 1840. As might be expected, the 1955 activity came as a complete surprise to the Puna people, comparable to the shock of the Hilo people in April 1946 who saw the waterfront washed away in the first large tsunami to hit the town in 69 years. There was even more surprise 14 years later in 1960 when another tsunami again destroyed the rebuilt waterfront. With the lesson driven home, the waterfront was made parkland and rebuilding disallowed. In Puna, the 1955 flows cut off most roads in the area and destroyed 15 houses in the town of Kapoho located just north of the Kii flow. However, the sigh of relief in Kapoho when the eruption stopped was short lived, and just five years later in 1960 the entire town was lost to lava. Kapoho in Hawaiian means the sunken place, and it is a fitting description of the area located between two fractures on the Kilauea East Rift. The land between the fractures had dropped down, creating a small low flat valley with a warm freshwater pool nearby supplied from seepage out of one of the enclosing fractures. It was a beautiful spot, but perfectly sited to collect any lava flow that happened to appear nearby. The year of 1960 was one to remember with Kapoho destroyed by lava in January, and the Hilo waterfront by tsunami in May. The 1960 lava flow is shown on the extreme upper right just north of the Kii flow. At Kapoho an attempt was made to divert the lava by throwing up earth retaining walls across the front of the flow, but other vents opened up and all walls were eventually breached.

The overall pattern of the Kilauea lava flows shows many of the same characteristics as the fracture system patterns. The vents where lava erupts and the flow begins tend to lie on directionals. This can be seen most clearly in the Puna area, where a number of the flows appear to be contained within directional boxes, with the south flows enclosed by N40W and N50E directionals, and the north flows by N50E and N7E directionals. On the east the Kii flow trends W2N, along with the base of the 1960 flow lying just above. On the left of the figure the Southwest Rift flows also seem to fit within sets of directionals much like the flows in the Puna area. In the next figure attention will be focused on another lava flow area in the south of Hawaii Island.

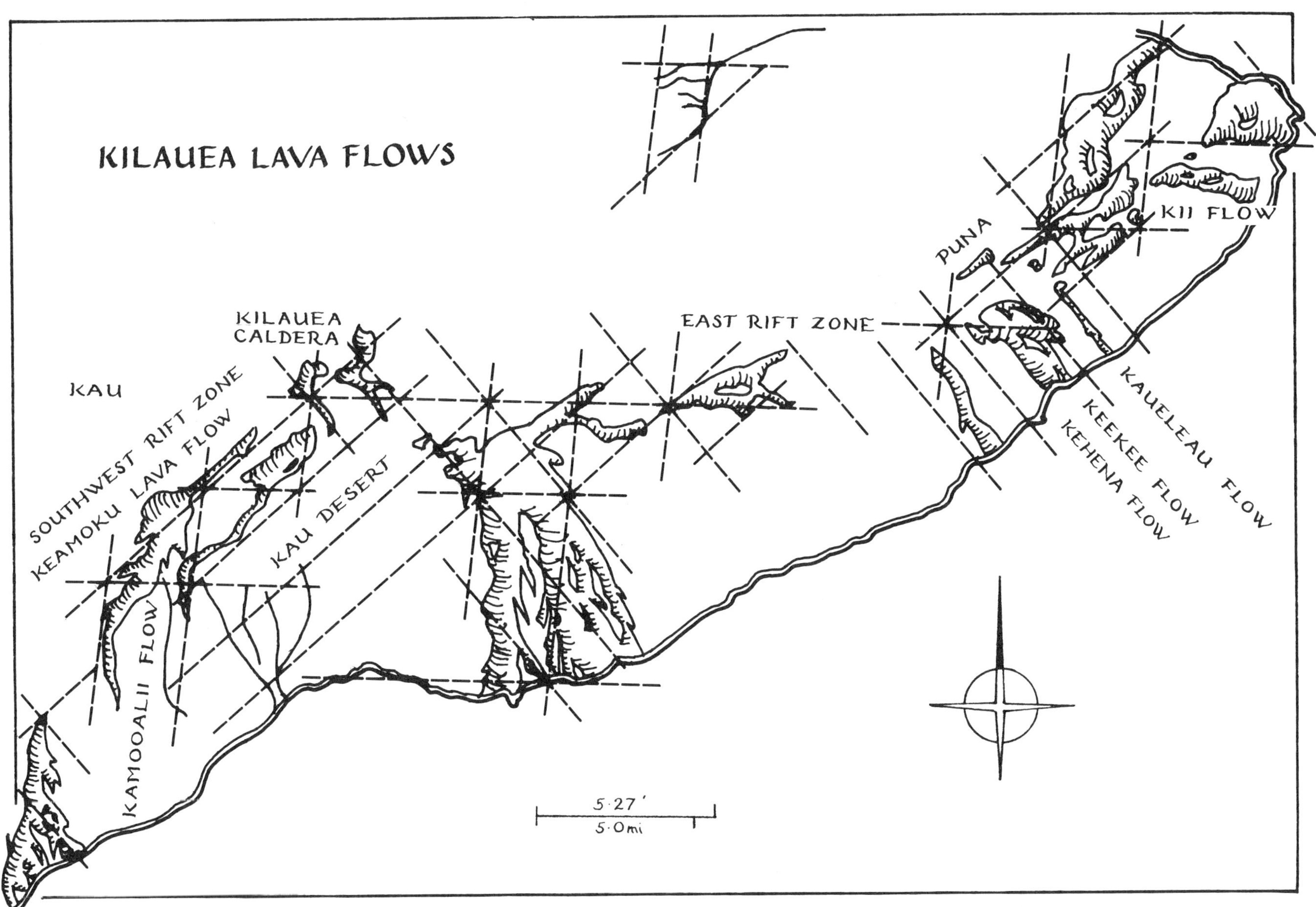
KILAUEA LAVA FLOWS
KAU
KILAUEA CALDERA
SOUTHWEST RIFT ZONE
KEAMOKU LAVA FLOW
KAU DESERT
KAMOOALII FLOW
EAST RIFT ZONE
PUNA
KII FLOW
KAUELEAU FLOW
KEEKEE FLOW
KEHENA FLOW
5.27'
5.0mi

3-9 *Mauna Loa Lava Flows*

This shows lava flows from the Mauna Loa Southwest Rift Zone. The rift begins at the top center of the figure and trends N50E for a short distance before shifting to a N7E directional that it follows most of the way to the ocean in the south. The flows in the lower half of the figure from the late 1800's and early 1900's do not generally trend along directionals, but rather are constrained within the directional pattern. These southern flows can be divided into three major streams that are separated approximately by hierarchical 2.64 min intervals. On the west is a single tongue that came from a 1907 flow, and in the middle are two tongues with the tongue on the left from the same 1907 flow and the right tongue from a 1887 eruption. The third more complex stream on the east occurred in 1868. The N7E trending west coast is intersected by five lava streams from the Southwest rift. They initially move N50E down the rift for some distance before shifting to W2N as they come down the west flank and into the ocean. The lower flow called Hoopuloa occurred in 1926, and the Alika flow just above came from a 1919 eruption. The three upper flows were fed by a June 1950 eruption that was one of the largest lava flows ever produced in Hawaii. In only three weeks more than 400 million cubic yards of lava spewed out to cover a 35 square mile area. The only previous Mauna Loa eruption to equal this occurred in 1859, but it took ten months, about 20 times as long, to pump out the same amount of lava.

The eruption began June 1, 1950 at the 12,000 foot level on the Southwest Rift with a two mile continuous line of fountaining lava called a curtain of fire. A short time later at the 8,000 foot level another eight mile long curtain of fire began, fountaining up to 1,000 feet high. Lava poured out from these fissures and flowed rapidly down the west flank of Mauna Loa at about six miles an hour, with the first tongue reaching the sea early the next morning and the next two tongues on the following day. During this first one and a half days an estimated 200 million cubic yards of lava was produced, representing half the total three week flow. In this huge flow of lava several buildings were destroyed but no one was injured. This is usual with Hawaiian eruptions that generally occur in lightly populated areas with little violence, and produce slow moving lava flows that give the people plenty of time to move out of the way. The main problem of the authorities during an eruption is keeping away the overly curious and careless visitors who have no idea of the risk involved.

The western flank of Mauna Loa is relatively steep, and this contributed to the faster than average flow of lava down that slope in 1950. This flank also has two major fault zones that run just inland along the west coast. These are the Kealakekua Fault System which is located just north of the figure, and the Kaholo Fault Zone that lies about 15 miles inland and crosses the west flank lava flows shown in the figure. These fault zones outline an area where part of the west flank has broken away and slumped down into the ocean in much the same way the east flank of Kilauea has shifted out and dropped down. Underground lava moving through the Southwest Rift of Mauna Loa continues to add material laterally to the rift, widening the zone. This tends to shift the area west of the rift towards the sea, with the flank sloughing off and slumping down into the ocean from time to time as it becomes top-heavy.

The correspondence between the Mauna Loa lava flows and the directional grid is much the same as found for the Kilauea flows. Although some flows follow directionals, a number do not, but rather tend to begin, change course, or end on directionals within the rhombus strain pattern, indicating that the overall grain of the land controlling the lava flow reflects the directional grid of strain lines representing preferential flow paths.

The streams shown on the east flank also appear to be constrained by the directional grid. It may be recalled that this is the only place where the volcano Ninole was not covered by later lava flows from Mauna Loa, so that stream erosion could continue for many thousands of years. The streams cut out valleys and shaped the valley floors into gently sloping surfaces, leaving behind the high remnant hills between the streams. Behind the hills the streams begin on a rather steep slope that grades up into the more recent lava shield of Mauna Loa. In 1868 this area was drenched with heavy rains and then struck with a major earthquake, bringing down a huge mud slide of water soaked soil that buried a village of some 30 people. The general orientation of the east flank stream system is N40W with many of the individual streams following the trend. The system is bounded on the east by a N7E directional, and on the southeast by a N50E directional that parallels the coast. In the northwest the stream headwaters line up approximately along N50E directionals. These boundary directionals serve to box in and constrict the system. In the next figure, a second stream system on Maui will be examined.

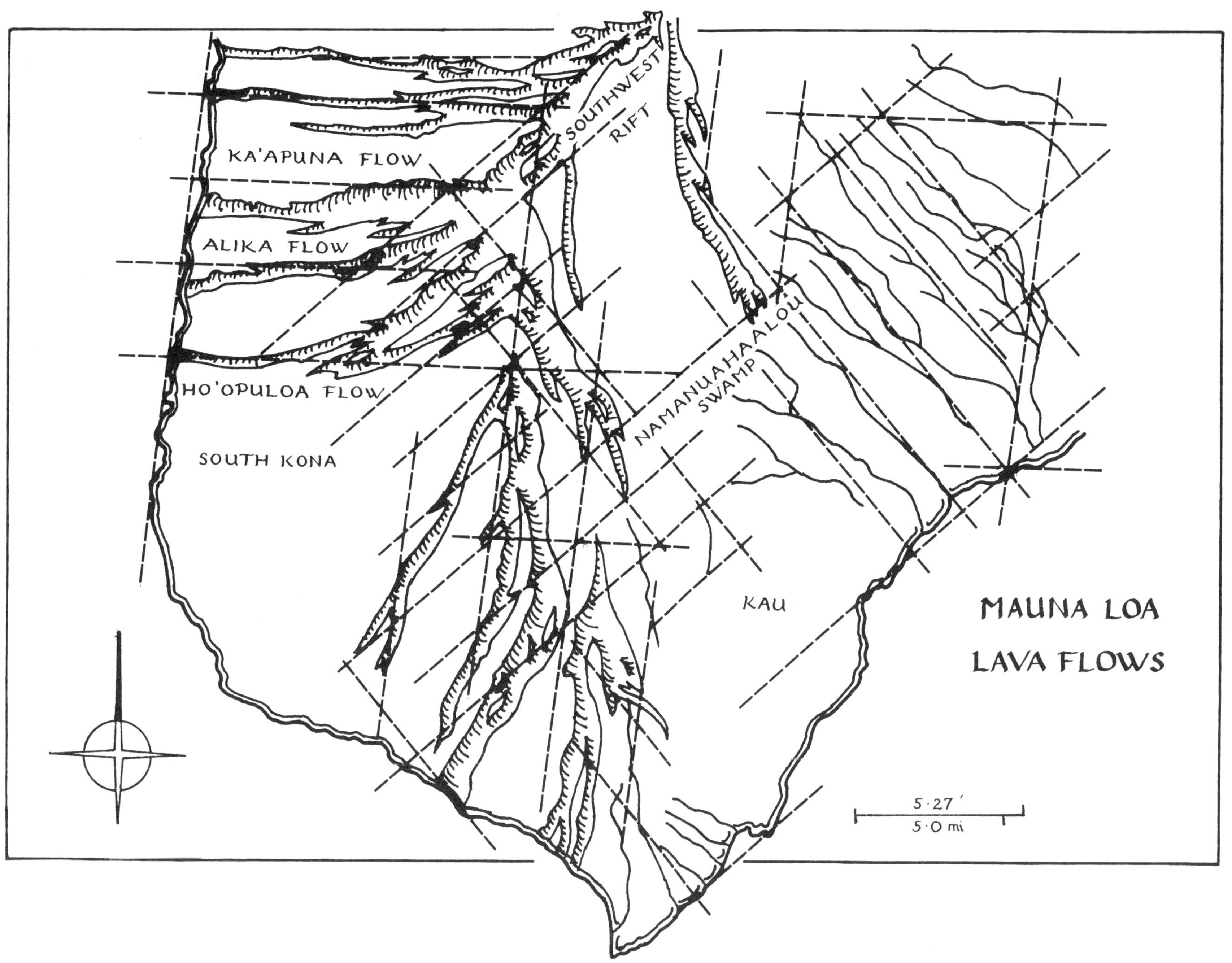
KA'APUNA FLOW
ALIKA FLOW
HO'OPULOA FLOW
SOUTH KONA
SOUTHWEST RIFT
NAMANUAHAALOU SWAMP
KAU
MAUNA LOA
LAVA FLOWS
5·27'
5·0 mi

3-10 Maui Streams

Shown here is the eastern portion of Maui with its streams and historical lava flow on the southwest coast that occurred around 1790, about the same time as Kilauea on Hawaii Island was exceptionally active. The date of the eruption was estimated by comparing two maps from two early explorers. In 1786 the French explorer La Perouse mapped a large bay at the eruption site which he appropriately named La Perouse Bay. Seven years later in 1793 the English explorer Vancouver surveyed the same area and found a smooth coastline and no bay, indicating that an eruption must have occurred around 1790 that filled the bay with lava. East Maui is essentially Haleakala Volcano, the highest of an interconnected group of six volcanoes in the immediate area that include two each on Maui and Molokai, plus Lanai and Kahoolawe. Several times in the past the sea level has dropped, and these four islands with their six volcanoes became one large island about half the size of Hawaii Island.

The Haleakala Volcano with an elevation of 10,000 feet is the third highest in the Hawaiian Islands. At the summit is a desolate crater of about 15 square miles filled with a number of multicolored volcanic cones that change color with the rising and setting sun, and it was here that the great Polynesian Hero-God Maui caught the Sun in his snare. Maui is known throughout the Pacific as a playful, mischievous, and crafty hero of many wonderful adventures. There is a possibility that he is the same sky figure as the constellation Hercules, who circles the North Pole with his left foot on the dragon's head while resting on his right knee, which marked the North Pole site about 9,000 years ago. On windward East Maui near Kailua, Maui's right knee can still be seen where he stopped to drink at a stream. His mother was the great fire goddess Hina who formed him by herself without anyone's help. The Hawaiian Islands, as well as New Zealand far to the south, were pulled up out of the sea by Maui using his great star

fishhook known as Scorpio. He also pushed up the sky so that mankind could stand up and walk on two feet instead of crawling on all fours. Like Prometheus, Maui brought fire to mankind, but only after he had been tricked by the mudhens who scratched out the fire and told him that the taro stalk and ti leaf held the secret. After Maui rubbed the plants together and found only empty hollow stems, he finally made the mudhens tell him that fire was truly in the tree of life growing in the sacred water. At one time he shook the whole foundation and framework of the Earth and brought down the great God Kane from the high heavens above and Kaneloa from the waters beneath the Earth. This may have happened when he knelt at the North Pole and turned the Earth, perhaps causing the framework to shift and the Sun to speed up and move so fast that his mother Hina could not dry her bark cloth. So Maui took his trap lines up to Haleakala, the house of the Sun, and caught the Sun while sitting under his grandmother's big Wiliwili tree, forcing the Sun not only to slow down, but also to make the summer days longer so the bark cloth would dry more easily.

In contrast to most volcanic craters that form by collapse as lava flows out from below, Haleakala is mostly erosional. Many years ago volcanic activity on Maui stopped for awhile and allowed streams to cut deep valleys into the mountain, especially on the wet windward north and east slopes. As erosion continued, the valleys extended upslope and eventually broke through the walls of the summit crater to merge and form a large erosional depression that stretched across the mountain top. When volcanic activity started again, rifts opened up in the depression and lava poured out to flatten the floor by filling the lower spots, and then to flood out and fill the valleys as it flowed down the mountain to the sea. Lastly, volcanic cones formed around the vents on the flat lava floor of the depression, completing the crater as it is seen today with its sculptured hills of many colors scattered across the walled enclosure.

The directionals that outline Haleakala Crater are well defined, connecting the stream headwaters on all sides to form an inverted "V." The north apex of the "V" at Koolau Gap is the right angle intersection of a N40W directional on the right and a N50E on the left that can be extended to the southwest to connect with the 1790 lava flow. The southern border of the "V" parallels the northern border across a 5.27 min interval, and is again formed by a N40W on the left and a N50E on the right that follows Kaupo Gap, formed when lava broke through the summit crater wall and flowed down the mountain. The stream headwaters on the south can also be joined along a W2N directional. Although individual streams often diverge from directionals, the overall grain of the stream patterns appears constrained by the directional pattern with stream branches, breaks, and course changes often occurring on directional grid lines. South of the crater the stream grain generally follows N40W around Kaupo Gap and Kipahulu Valley, but then shifts rather abruptly to W2N on the east. Abrupt stream pattern shifts of this type can also be observed on the windward slopes of Mauna Kea on Hawaii Island, and this will be discussed on the next page.

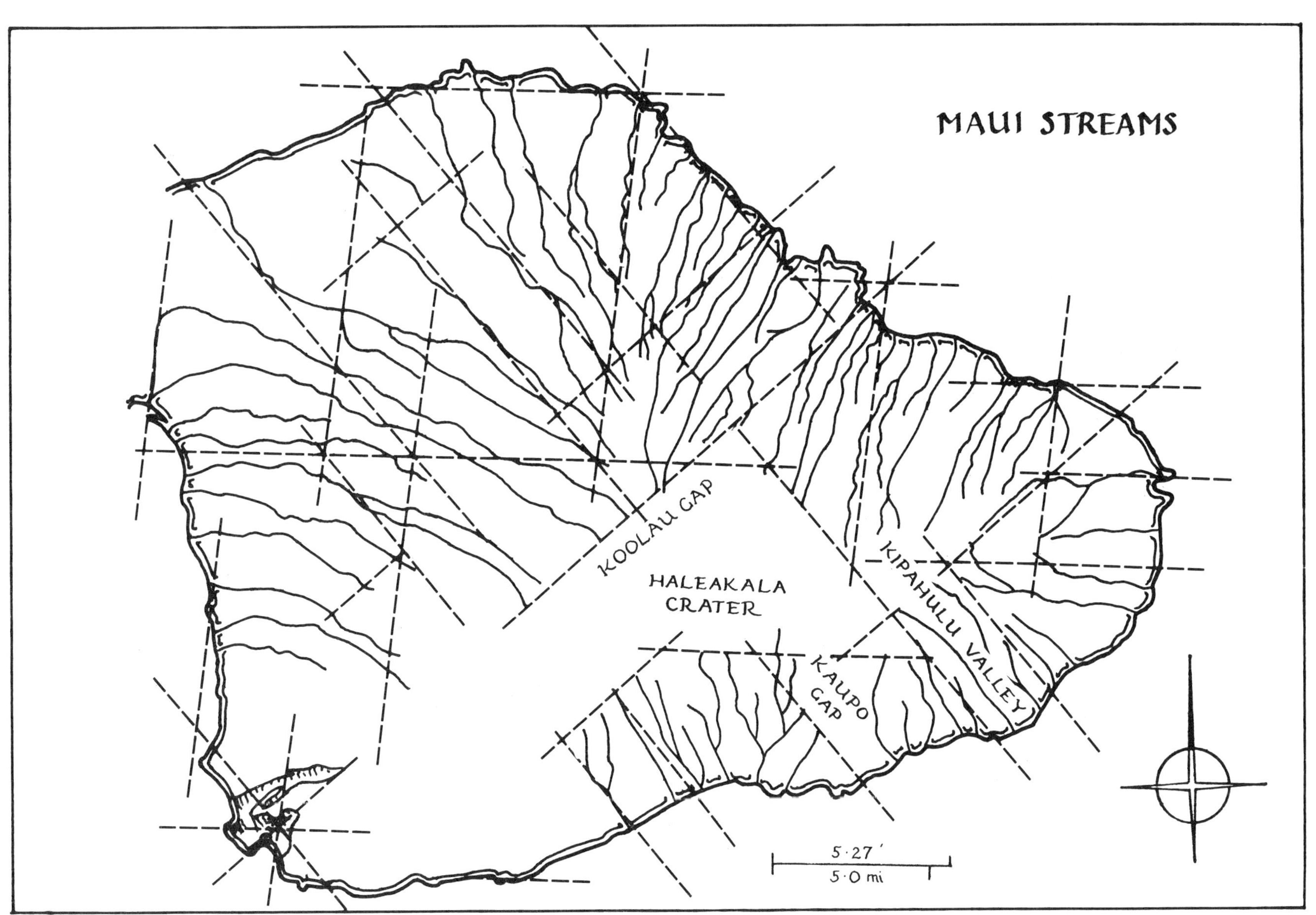
MAUI STREAMS
KOOLAU GAP
HALEAKALA
CRATER
KIPAHULU VALLEY
KAUPO GAP
5·27′
5·0 mi

3-11 Hawaii Streams

This shows the radial stream system, resembling the spokes of a wheel, on the north coast of the Island of Hawaii. This Hamakua coast runs from Hilo in the lower right to Honokaa in the upper left, and is the windward side of the island exposed to the moist tradewinds blowing in from the northeast. As the wet air is forced up the mountain slope of Mauna Kea and cools at the higher elevations, much of the moisture drops out as rainfall that can reach 300 inches annually. The streams carrying the run-off down the mountain side to the ocean have cut deep valleys that are choked with thick rainforest vegetation. On the flat land between the valleys sugar cane is grown up to an elevation of about 3,000 feet, thriving in the wet warm climate. However, with the recent drop in sugar price, cane fields have been abandoned and sometimes replanted in even rows of macadamia nut trees, and the rectangular patterned orchards make a pleasant contrast with the green expanse of the cane fields.

Further up the slopes of Mauna Kea the fields of sugar cane grade into woodlands, then into shrubs and grasses, and finally above the tree line, into a barren expanse covered with lava and cinders from past eruptions. With the last eruption occurring less than 4,000 years ago, Mauna Kea is considered dormant rather than extinct, so that volcanic activity could resume at any time, although the risk is relatively small. Viewed from Hilo at sea level, the top of Mauna Kea appears bumpy with several white dots perched on top of the bumps. The bumps are old cinder cones from past eruptions, and the dots are astronomical observatories situated to take advantage of what may be the best telescope seeing conditions in the world. Although clouds are common around the mountain, they almost always lie below the summit, giving clear nights and excellent seeing more than 80% of the time. The mountain is also surrounded by the large expanse of the flat blue Pacific Ocean so that airflow at the summit is relatively smooth and even with little turbulence. In contrast, a continental land mass is subject to considerable differences in heating from the Sun due to variations in elevation and the various colors of rocks and vegetation, causing telescope images to shimmer from the shifting and turbulent winds. In addition, Hawaii Island has little industrial activity and no large cities, with a total population of only about 100,000, and with only 40,000 people living in Hilo, its largest city. Therefore, there is little pollution from factory smoke, car exhaust, and city lights, making Mauna Kea one of the darkest observation sites on Earth. But Hilo is large enough to provide the material necessities of good housing, communications, transportation, and other facilities for day to day operations.

Another special advantage that is very important for infrared observation is the lack of humidity and rainfall at the summit of Mauna Kea which rises above the region of heavy rain where most of the moisture from the tradewinds is dropped. Infrared radiation is essentially heat coming from a portion of the spectrum outside the visible light range. Most people are intimately familiar with infrared rays which cause the nasty sunburns from a day on the beach. Any moisture in the air tends to soak up the infrared radiation before it reaches Earth, so that the very dry summit of Mauna Kea makes it a choice location for the special infrared telescopes that measure the variations in heat coming from objects in the sky. Of the seven telescopes operating or under construction, four are specifically designed for infrared viewing. The United Kingdom 150 inch and the NASA 120 inch have been in operation for some time. Two other new telescopes operating in the submillimeter range, or extreme infrared, are the 590 inch James Clerk Maxwell of the United Kingdom, and the 400 inch Kresge of Caltec. Other instruments on Mauna Kea that collect visible light from the stars are the 88 inch University of Hawaii and the 142 inch Canada-France telescopes. The 400 inch Keck telescope now under construction will be the largest optical telescope in the world. It will have a light gathering surface made of 36 mirrors, each of six sides, that will be adjusted twice each second, so that all the reflected light will converge to the same point and form a single composite image.

The radial stream pattern downslope from the observatories shows much the same characteristics as that on Maui. Starting at the lower right, the grain of the stream pattern just above the Wailuku River and west of Hilo Bay trends W2N. Moving counterclockwise, an abrupt change to N50E occurs where the coast breaks back and begins to curve around to the northwest. This trend is maintained until it again changes fairly sharply to N7E about halfway up the coast. Although individual streams tend to wander, the overall grain within each of the segments appears constant, giving the impression of rather sharply divided intervals of coastline within which a directional trend is maintained. the tendency for stream headwaters to line up along directionals can also be seen. A good example of this is the rather empty 5.27 min interval between two N50E directionals about halfway up the coast. In contrast to this radial stream system are the parallel stream systems of Oahu which will be examined next.

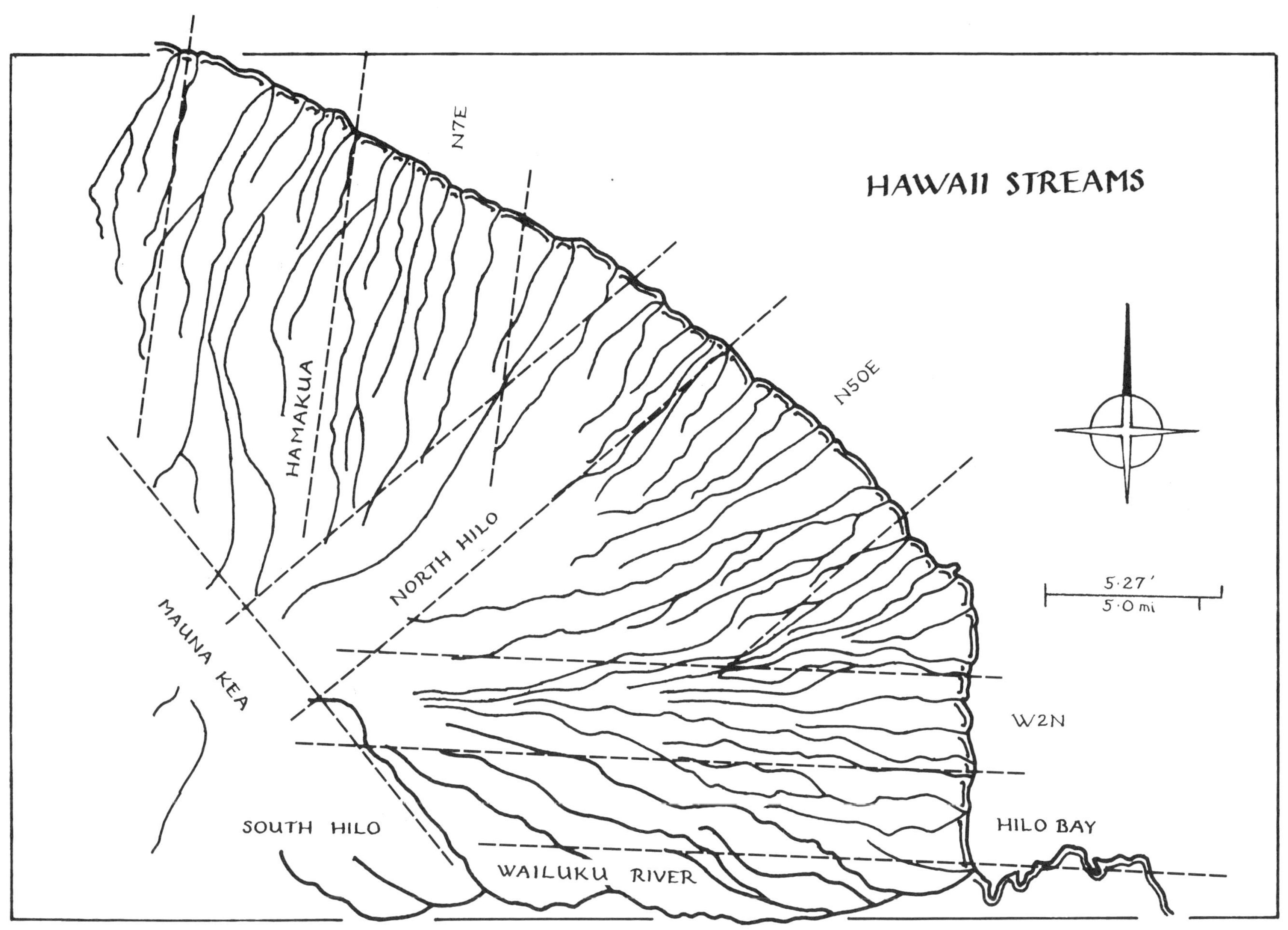
HAWAII STREAMS
N7E
N5OE
HAMAKUA
NORTH HILO
MAUNA KEA
W2N
SOUTH HILO
HILO BAY
WAILUKU RIVER
5·27'
5·0 mi

3-12 Oahu Streams

This shows the coastlines and stream courses on Oahu Island where Honolulu is located. The island is formed of two elongated volcanoes running parallel to each other along N40W trends. These volcanoes have been inactive for thousands of years, allowing streams to cut out deep valleys and severely erode the mountain ranges. The large northeast portion of the island is a rectangular block formed of the Koolau mountain Range and bordered by right angle intersecting N40W and N50E directionals. There is a major inset at Kaneohe Bay on the northeast coast, that is reflected in an offset of 2.54 min along the N40W crest of the Koolau Range. Along the northeast flank and the southern half of the southwest flank of the Koolau Mountains, the streams flow down the slopes perpendicular to the crest, forming a pattern of parallel streams with the grain trending N50E. However, in the upper left of the Koolau block, the streams come down off the crest along a W2N trend that can be extended further west along the north coastline of the Waianae Mountain Range. This mountain, forming the western portion of Oahu, is framed by a parallelogram with the longer east and west sides following N40W, and the shorter north and south sides trending W2N. The crest of the Waianae Range follows N40W in the north, but makes an abrupt shift to N7E in the southern part of the parallelogram. The streams in this southern portion flow off the N7E crest and run towards Pearl Harbor, although most disappear before they reach the coast. This is not uncommon for Hawaiian streams that flow over highly porous and fractured lava rock which soaks up much of the water carried by the streams. The headwaters and ends of these southern streams lie on two N7E directionals separated by a 2.64 min hierarchical interval.

The high porosity of the lava rocks in Hawaii is created by the release of volcanic gases as the magma flows up from deep in the Earth to the surface. Any liquid that lies deep underground contains dissolved gas held in solution by the high pressure. However, as the liquid flows up to the surface the pressure continually decreases, allowing bubbles of dissolved gas to escape. Below the surface the spaces created by the gas bubbles are quickly filled in by the liquid magma, but when the magma pours out on the surface as lava, it rapidly thickens and hardens, and the bubble spaces remain in the rock making it highly porous. In addition, there are often spaces left between the thin layers of lava that are piled up on top of each other, while other openings are formed from cracks along fracture zones. The end product is a leaky bed of lava rock that will hold water only a short time. When rain falls on ground made of this leaky rock, the water tends to flow down through the rock rather than collecting on the surface, so that many streams disappear into the ground over a fairly short distance and never reach the sea. This means that streams, lakes, and ponds are rather undependable sources of water supply, so that other types of natural water reservoirs must be utilized.

Rain that falls on the porous lava rock and soaks down into the ground gradually percolates down through the layers of lava until it reaches sea level under the island where salty ocean water has filled the porous rock. Luckily however, the fresh rainwater is lighter than the salt water from the sea, so that it tends to float on top of the sea water, much like cream floats on top of milk. As the fresh rainwater continues to filter down through the rock, it accumulates just above the saltwater in a saucerlike lens with the saucer edge following the coastline of the island. Much of Hawaii's water supply comes from large wells called Maui wells drilled into this freshwater lens. An inclined shaft is driven down to sea level and a large chamber excavated that stays full of freshwater by seepage through the porous rock walls. These wells supply enormous quantities of water with capacities of up to 50 million gallons each day. However, the source is limited and has proven inadequate to provide sufficient water for the large and growing population on Oahu, that has about 10% of the Hawaiian land area and about 80% of the total population of about one million people.

Another large source of fresh water in Hawaii is contained in porous lava rocks at high elevations that have been sealed off by non-porous walls of solid volcanic basalt. It will be recalled that many dikes form underground when magma flows into fractures and hardens. Under these conditions much of the gas stays in solution rather than bubbling out, so that the magma solidifies as a solid thin wall of rock. These dike walls form the sides of natural water reservoirs that often hold very large quantities of fresh water collected over many years. The reservoirs are tapped with Lanai-type wells that are drilled horizontally into mountain sides and through the solid dike walls to reach the water-filled porous rock within. In some places fresh water also collects above thin horizontal sheets of dense material such as volcanic ash or clay, but volume in these reservoirs is usually much less than in the saucerlike lenses below the islands and the dike reservoirs in the mountains. On the next page the larger streams of Kauai Island will be studied. Kauai is geologically the oldest of the major islands, so that there has been more time for streams to cut into the land and erode out patterns.

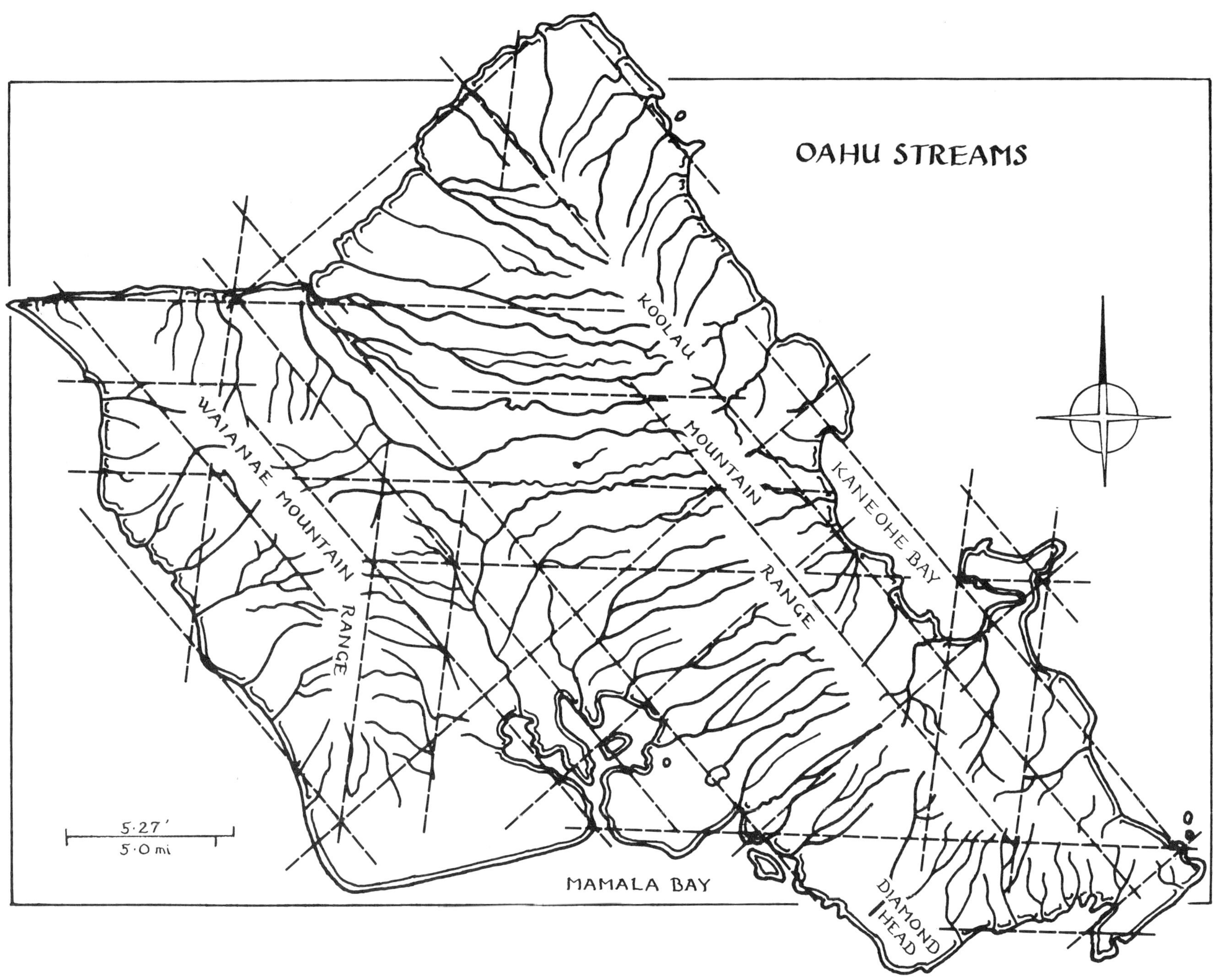
OAHU STREAMS
KOOLAU
WAIANAE MOUNTAIN RANGE
MOUNTAIN RANGE
KANEOHE BAY
MAMALA BAY
DIAMOND HEAD
5·27'
5·0 mi

3-13 Kauai Rivers

This shows the coastline and river systems of Kauai Island. These rivers are merely larger streams than generally found in Hawaii and would not be called rivers in other parts of the world. However, Kauai is the only island of the chain where streams have had the time to develop an extended system of tributaries that form a distinct pattern for study. The stream patterns can be divided into a number of pie-shaped sectors formed by four directionals that intersect in a node at about the center of the island. The two diagonal directionals passing through the node follow the Alakai Swamp along N40W and the Honopahu Ridge along N50E. They form an "X" that holds the four streams, Wainiha, Lumahai, Hanalei, and Kalihiwai between the two upper cross bars, and the Waimea Stream system between the cross bars on the west. A well defined N7E directional also intersects the node and divides the Hanalei, Kalihiwai, and Wailua systems on the east from the other four stream systems on the west. Consistent with the strain pattern analogy, the W2N directional passing through the node is not as clearly defined, but can be followed for a short distance west of the node along several Waimea tributary beginnings.

The central node is the location of the Mount Waialeale summit that rises to more than 5,000 feet and marks the eastern edge of what was once a large caldera some 13 miles across. About four million years ago in the area now occupied by the Wainiha and Lumahai stream systems, the caldera was a depression on top of a volcano filled with massive lavas that had cooled slowly and were much harder than the thin lava layers that formed the rim. Since that time, streams have eroded away much of the softer rim, cutting out 3,000 foot valleys, but leaving much of the harder lavas that made up the broad flat floor of the caldera. This is now a high plateau, partly covered by the Alakai Swamp where rainfall of at times more than 50 feet a year has given it the honor being the rainiest place on Earth.

The streams in the caldera area are enclosed in a rectangular box with sides made of two pairs of N50E and N40W directionals. The southeast side follows the N50E directional that forms one of the intersecting diagonals making up the central cross, and a parallel N50E forms the side on the northwest. The two shorter sides of the rectangle are composed of two N40W directionals, with an additional five N40W directionals between them separated by 2.64 min intervals, forming a set of six rectangular strips that divide up the stream systems. The two strips on the lower left contain the Waimea stream system which is separated from the Wainiha system to the northeast by the third strip that encloses the Alakai Swamp. The last two strips hold the Lumahai stream system which is bounded on the east by the N7E directional that intersects the node. The large rectangle containing the stream systems northwest of the node also has a smaller rectangular appendage on the southwest that outlines the lower courses of the three main Waimea stream channels that trend N50E on their way to Waimea Bay and the sea.

It was at Waimea Bay that Captain Cook first stepped ashore in the Hawaiian Islands on his way north in search of a Northwest Passage above North America that would connect the Atlantic with the Pacific. Ships sailing from Europe to the Orient had to travel far to the south around South Africa or South America, so that the discovery of a Northwest Passage would shorten the trip considerably. The First Lord of the Admiralty, the Earl of Sandwich, put Captain Cook in charge of two ships, the Resolution and Discovery, that sailed from England in search of a Northwest Passage in 1776, the year of the United States Declaration of Independence. His course took him around the Cape of Good Hope on the southern tip of Africa, across the Indian Ocean to the South Pacific, and on to Tahiti. He had made two earlier voyages to the South Pacific in 1772 and 1775, when he had explored south of the equator searching for the huge mythical continent of Terra Australis. On these trips he mapped the coast of New Zealand and part of Australia, discovered many small islands, and crossed the Antarctic Circle three times to find nothing but icebergs, fog, and cold, rainy winds, but no Terra Australis. However, in January 1778 he was sailing north from Tahiti bound for the north Pacific when the crew spotted Tropic Birds and Boobies flying above and turtles swimming in the sea, and then on January 18, 1778 the island of Oahu was seen to the northeast, and a little later, Kauai to the north where he landed and stayed two weeks to provision his ships. He named the new land the Sandwich Islands in honor of the Lord of the Admiralty before sailing on north to look for the Northwest Passage. A year later after an unsuccessful search, he returned to map the islands for the first time, and then stopped to provision at Kealakekua Bay on the western side of Hawaii Island where all went well and he left without incident. Unfortunately, a mast was broken in a storm, and when he returned to Kealakekua he found the natives had grown decidedly unfriendly. And it was there that Captain Cook, one of the world's greatest explorers, was killed in a argument over a stolen boat.

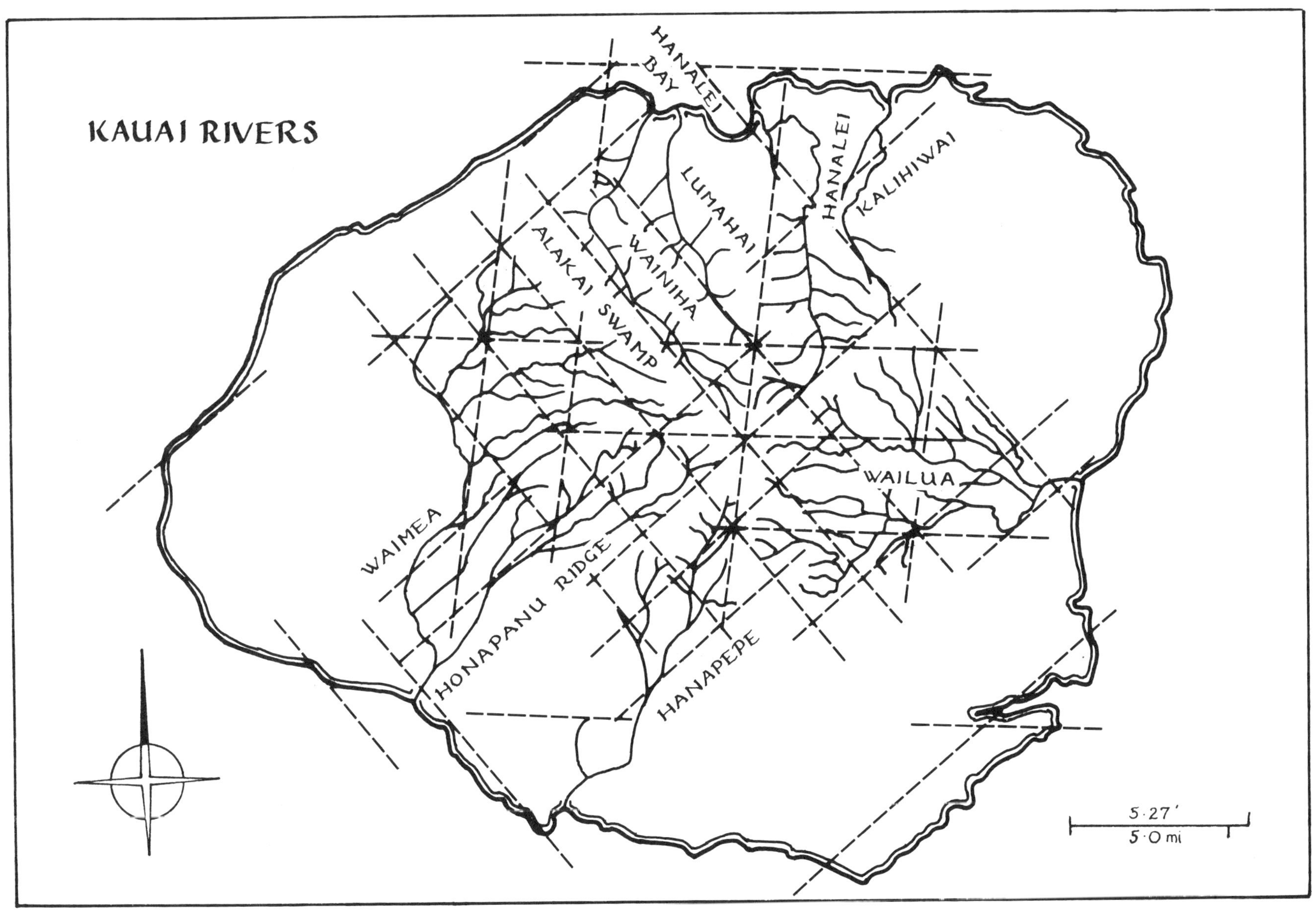

KAUAI RIVERS
HANALEI BAY
HANALEI
LUMAHAI
WAINIHA
HANALEI
KALIHIWAI
ALAKAI SWAMP
WAIMEA
HONAPANU RIDGE
HANAPEPE
WAILUA
5·27'
5·0 mi

In this chapter a small area of the Earth was selected for study to determine by example how certain geological features might correspond to the strain pattern analogy. Based on the analogy, it was suggested that the directionals within the strain pattern represented paths of preferential flow, both on and within the Earth's crust. These preferential flow paths carrying fluid might gradually clog up as solids dropped out or mineralization occurred, or they might be opened up as material was scoured out or dissolved away. The degree to which clogging or opening up happened would depend on the type of fluid and channel material, velocity of the flowing fluid, type of solid particles being carried, and changes in temperature and pressure, among other things. However, whatever the circumstances, the strain pattern with its net of directionals would act as a constraining framework, redistributing fluid flow to create a grain within the flow system corresponding in some way with the strain pattern.

The Hawaiian Islands were selected for study because they have extensive shorelines, recent lava flows, many streams, and a wide range of fracture systems located around the islands. This provided a large amount of study material in a relatively small area that could be used to establish characteristics of correspondence between geologic system features and the strain pattern. Although individual fractures do not generally follow directionals, it was found that they tend to begin, change direction, branch, break, and end on the strain pattern directionals. This is consistent with the mechanical strain analogy where a break in a stressed material rarely occurs along a line of extreme strain, but rather along imperfections in the material. The strain pattern seemed to serve as a supporting framework that tied together the discontinuities in a fracture system, producing a directional pattern grain with an orientation often considerably different from that of the individual fractures. This fracture system correspondence was also observed to a lesser degree in lava flows and stream systems, but more often these flow systems had a

linear grain that tended to follow a particular directional. This was clearly defined in the radial stream system on the Hamakua coast of Hawaii Island. The system was divided into a number of distinct sectors with the streams in each sector orientated along one of the four directionals, and with the directional change quite abrupt between adjacent sectors. On Kauai the river system boundaries were very clearly defined by the directional grid with a number of the individual systems sharply separated across hierarchical intervals. For the relatively small area of the Hawaiian Islands, the strain pattern analogy seems to hold up reasonably well.

However, a geologic analysis is never complete because there is never an end of new material from nature to be integrated with an explanation, and old material can always be rearranged and examined from an endless variety of formats and patterns. In addition, the testing process itself can take many forms that include simple visual checks, statistical checks, or formal model checks, with the testing designed to take a broad look at many items or a detailed look at a very few selected items. There is no right way to test an explanation, but rather a multitude of possibilities, each of which may provide some unique additive to an explanation. With each approach there are pitfalls and problems that serve to bias the test. It is often said that statistics can prove anything, and essentially that represents a simple statement of the selectivity and interpretation problem that is so dominate in geologic analysis. Statistical testing is also rigidly constrained to the study of numerical quantities, and this sharply limits those aspects of nature that may be selected. Often the large non-numerical portion of nature that drops through the statistical sieve is of greater importance than the residual numbered portion retained. This problem of numerical limitation also occurs in formal model testing, that basically is a mathematical extension of the statistical method. Visual testing also has problems related to selection and interpretation. Selection of examples for comparison from

infinite nature is always subject to the unconscious bias of the person involved. In addition, no simple manmade pattern can possibly duplicate the complicated intricacies of nature, so that certain aspects must always be passed over or ignored while others are emphasized.

The analysis problems discussed above may be minimized in two ways, one of which is to restrict the study to a limited set of data and apply a variety of testing methods to the material. Although this approach is rarely used in any one study, it is commonly used by the supporters or critics of a particular concept to reexamine the original material. The acceptance or rejection of new ideas often revolves around the lengthy process of comparing results from different analysis methods. The second alternative is to continue the search for additional examples that may add weight to the argument. This is the only viable approach to testing analogy by visual comparison and will be the approach used in this study. In the next chapter examples will be given from many parts of the world that help to reinforce the concepts of the correspondence between the strain pattern and Earth features.

4 THE EARTH

EXPANDING THE PATTERN

Much of geologic analysis is a study of examples. A pattern and its explanation may begin with a theoretical argument, but sooner or later comparisons with the physical world itself must be made. The degree of acceptance parallels the number of examples from nature that can be shown to be consistent with the new pattern. The examples chosen for comparison are a product of man's attempt to define and compartmentalize indivisible nature. The stated, or unstated, definitions used in an analysis essentially provide a set of constraints that serve to restrict the field of nature from which examples may be drawn, thereby reducing the number of possible examples. However, additional constraints are usually needed in order to further reduce the number of examples to a manageable size, and these could include restricting the area of study or perhaps the type of features studied. All these constraints provide a basis for a reasonable selection of examples from nature that may be analyzed without undue complication. Although this selection process is often subjective, unstated, and possibly unconscious, it may be a major factor in the development of study conclusions. A new geologic idea usually comes about by the mental selection of those items from nature that seem to fit together in an orderly way, and this process bares a close resemblance to the selection of examples to be used in consistency comparisons.

The selectivity process may be highly subjective and often may seem somewhat irrational. Every analyst is a flesh and blood human being whose mind has been shaped in a bounded environment by unique experiences that mesh with his innate mental facilities to form and reinforce a particular pattern. The experiences of childhood, school, and work are all absorbed and imbedded in the pattern and contribute to the end result. A child learns to speak and act within a family, developing a sense of communication based on concepts held by other family members. The emerging pattern serves to categorize experiences, taking into account time, place, and relationships, and providing a organized and meaningful place for all routine observations. New experiences are channeled through this processor pattern, and most are easily slotted into their respective niches to continually reinforce the knowledge and belief pattern of the individual. Experiences that do not quite fit can easily be unconsciously adjusted for consistency, and the residual that refuses to fit may just as easily be ignored. What does not fit does not exist, and therefore never really happened.

The personal preference aspect of selectivity in analysis is not a problem to be solved, but rather a fact to be acknowledged, and whether stated or unstated it exists to an extent in all studies. Increasing the number of comparison examples does not necessarily improve the analysis, because nature is an infinite spectrum that may be described and arranged in a multitude of ways, and the basic mental constraints that guide selectivity can almost always find more examples that conform to the processor pattern. What is generally required to improve the situation is a conscious shift in the constraints that determine the comparison examples to be studied. This shift is often forced on an analyst by an outside party, which is the method of the scientific community with their arguments and rebuttals that provide for the input of other individuals with different mental patterns and different ideas of selection constraints.

Thus far, two groups of examples from nature have been used to compare the directional pattern with natural features. In chapter 1 the examples from nature were severely restricted by form to relatively long sections of coastline, but with no area restriction so that examples could be selected from anywhere in the world. In the last chapter the restrictions were in effect reversed, with restrictions in features loosened to include systems of fractures, volcanic flows, and streams as well as coastlines, but with the area restricted to the Hawaiian Islands and the surrounding Pacific Ocean. From this restricted area set of examples it was found that the directional strain pattern developed in chapter 2 corresponded very closely with the patterns found in the various systems of natural features. The type of correspondence was also found to reflect in many ways the characteristics of the strain lines forming the directional pattern. As a further check on the strain pattern, a third set of examples can be chosen that include all features discussed in the last chapter, but with the area unrestricted to allow for examples from the world at large. This selection process will be used for the examples discussed in this chapter, but with the additional constraint that only natural systems that cover relatively large areas will be studied. This will serve to limit the example set to a reasonable number. The examples include an island group from the arctic and one from the temperate zone, a fracture system from the Indian Ocean and two others from continents on opposite sides of the world, both lava and pyroclastic volcano flows, and the three largest river systems in the world located in North America, South America, and Africa.

4-1 Queen Elizabeth Islands

This shows the Queen Elizabeth Islands that occupy the northern portion of the Arctic Archipelago lying above the North American continent. The whole archipelago complex is not only the largest island group in the world, but also contains some of the world's largest islands. It is split into two parts by the Parry Channel, with the Queen Elizabeth Islands to the north, and to the south, Baffin Island on the east and Victoria and Banks Islands on the west. The Queen Elizabeth islands lie just below the North Pole and are essentially a cold, barren, and windswept arctic desert with scattered drifts of snow and some ice caps on the eastern islands. The islands are virtually uninhabited, with a few Eskimo settlements on the southern fringe along Parry Channel and a few small research stations. The island group resembles a triangle with the north lobe of Ellesmere Island jutting out of the apex towards the northeast. The base of the triangle follows Parry Channel along a W2N directional from Melville Island on the west to Devon Island on the east, and the upper left side follows a N50E directional from Prince Patrick Island in the south up along the northwest coast of Ellesmere Island in the north. Only the southern part of the triangle's right side can be seen, following a N7E directional north from Devon Island in the southeast, until it becomes lost in Ellesmere Island where the east coast shifts to a N50E directional. The north lobe of Ellesmere Island has a N50E grain and is bordered on the southwest by a N40W directional that follows Nansen Sound and Fosheim Peninsula. The many islands south and west of Nansen Sound are separated by several channels and straits generally trending N40W that include Maclean Strait, Perry Strait, Austin Channel, and Bellantyne Strait. It is also noted that the ocean floor contours to the northwest follow N50E, parallel to the northwest border of the island group.

Although this cold arctic island system was virtually unexplored until the 19th century, there is evidence that the north lobe of Ellesmere Island has been occupied twice, the first time about 4,000 years ago at the end of the warm Altithermal period, and again about 700 years ago at the end of another warm period called the Medieval Optimum. Around 10,000 years ago the last ice age ended with the melting of the continental ice sheets in Eurasia and North America, and for the next few thousand years the weather was generally wetter and warmer than it is today. During this warm Altithermal period there were rivers and lakes across what are now the deserts of North Africa, The Middle East, and India, and cattle grazed there on the grasslands. The period also coincided with the beginnings and spread of virtually all aspects of civilization. In the Middle East between 10,000 and 6,000 years ago, animal herds and farming were established, town life began, and pottery, weaving, metal working, writing, and the wheel were discovered. Over the next couple of thousand years many of these changes spread up over Europe and out over Asia, and also were established on the North American continent. All over the world people moved north following the receding ice sheets, and eventually, as the permanent ice pack melted a few of the more hardy hunters reached the Arctic, along with the sea life. It was at this time that man settled in the northern part of Ellesmere Island until they were forced to leave about 4,000 years ago when the ice pack shifted south. However, about 1,000 years ago the warm Medieval Optimum period began and lasted about 300 years. It was at this time that the Vikings or Norsemen from Scandinavia burst out, not only to raid other communities, but also to establish their own settlements to the south in Russia, Brittany, and Sicily, and to the north in Iceland and Greenland. During this warm period Eskimo settlements also reappeared on Ellesmere Island, and recent evidence indicates that the Norse Greenlanders sailed north to make contact and trade with the Ellesmere natives. But with the close of the warm Medieval Optimum, the Little Ice Age appeared around A.D. 1500, and the Ellesmere settlements were abandoned, and the Norsemen in Greenland died out.

During the cold Little Ice Age glaciers advanced in the Alps, French grape harvests were delayed, and arctic exploration virtually ceased. Before the cold period began and the arctic pack ice moved south, Frobisher and John Davis had sailed to Baffin Bay just east of Devon Island, and Henry Hudson had explored the bay named after him. Then came the cold period, and it was over 200 years before the Little Ice Age ended and John Ross and Edward Parry sailed north in the early 19th century to find Melville, Bathurst, and Devon Islands. With continuing exploration the Northwest Passage was finally discovered by McClure in 1850, although it was not until 1906 that Amundsen sailed through the passage from the Atlantic to the Pacific around the top of North America. More recently, this arctic region has gained considerably in importance with the discovery of oil, gas, and minerals.

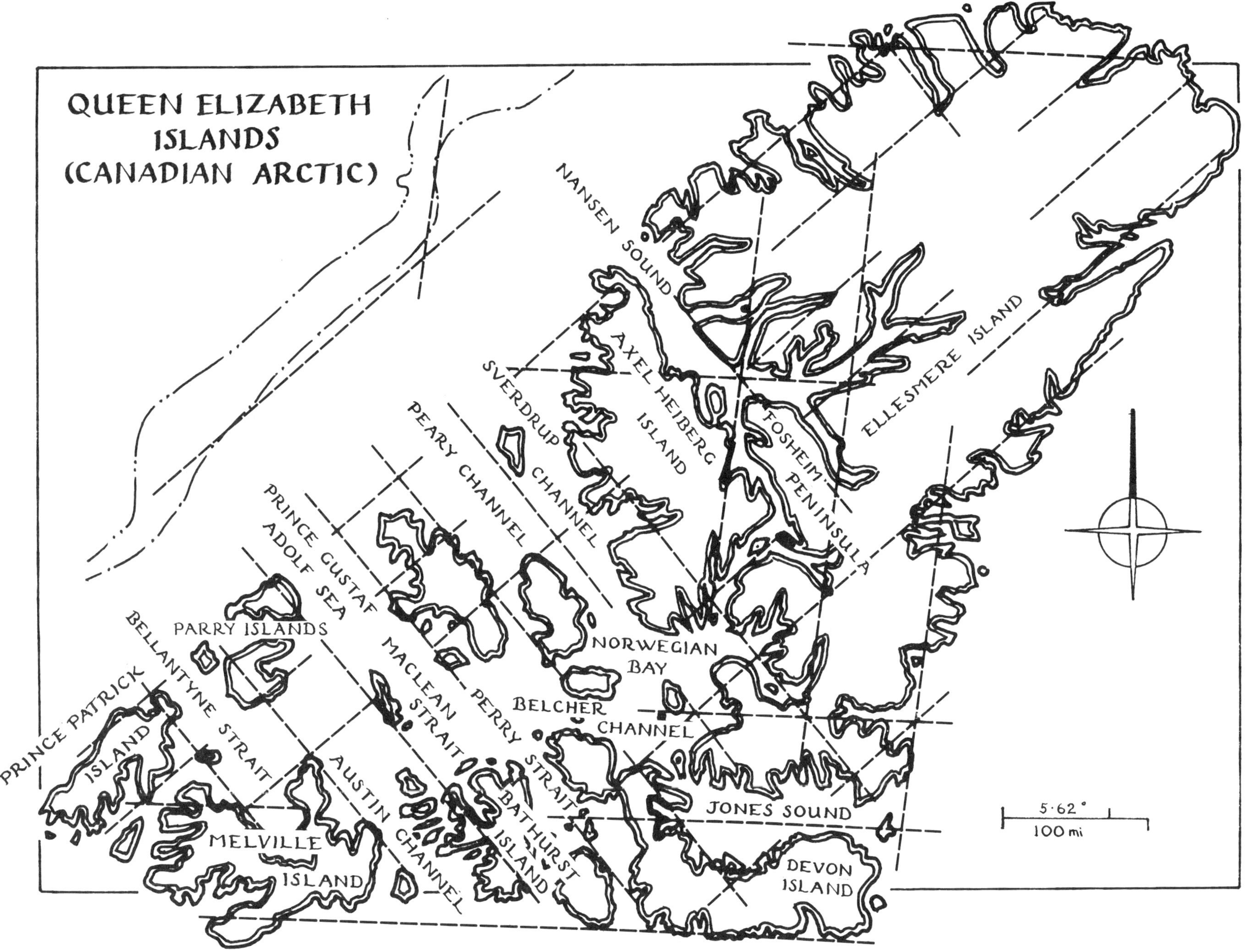
QUEEN ELIZABETH ISLANDS (CANADIAN ARCTIC)
NANSEN SOUND
AXEL HEIBERG ISLAND
SVERDRUP CHANNEL
FOSHEIM PENINSULA
ELLESMERE ISLAND
PEARY CHANNEL
PRINCE GUSTAF ADOLF SEA
PARRY ISLANDS
BELLANTYNE STRAIT
PRINCE PATRICK ISLAND
MACLEAN STRAIT
NORWEGIAN BAY
BELCHER CHANNEL
PERRY STRAIT
AUSTIN CHANNEL
BATHURST ISLAND
MELVILLE ISLAND
JONES SOUND
DEVON ISLAND
5·62°
100 mi

4-2 British Isles

This shows the coastline of the British Isles and the subsea contours to the west. The principal island on the east is divided into Scotland in the north and England in the south, with Wales jutting out into the Irish Sea towards the southwest. The small island groups of the Shetlands, Orkneys, and Hebrides north of the main island are all part of Scotland. West of the Irish Sea is Ireland which resembles an Irish sheep dog standing on end along a N7E directional, with the Republic of Ireland made up of the dog's body in the south plus its nose in the northwest, and Northern Ireland occupying most of its head in the northeast. West of the British Isles are the extensive fishing grounds of the Atlantic centered around the George Bligh Bank, Rockall Bank and Rise, and Porcupine Bank, which together show a general N50E grain. The Rockall Rise has a distinct rectangular shape with the sides oriented N50E and N40W, and crossed by a N7E directional that follows the southwestern edge of the Rockall Bank contour. These ocean features are separated from the British Isles by a N7E directional that shifts to N50E near the top of the page, following the shallowest undersea contours.

The main islands are divided up by a rectangular network of N50E and N40W directionals, along with several N7E directionals that follow other geographical features. A series of ten N50E directionals with 2.81 deg separations may be traced from north to south across the islands. the first follows the shallow underwater contours at the top of the page, and just below, the second follows the northwest coastline of the Outer Hebrides and Shetlands. The third N50E traces the Scottish coast along Moray Firth and the fourth traces the northwest coastline of Ireland, cutting off the nose and front paw of the Irish sheep dog. Continuing south, the fifth and sixth directionals outline the back leg of the Irish dog and the seventh follows the northwest coastline of Wales. The eighth directional traces the southern toe of England called

Cornwall and continues northeast to the Wash, while the ninth defines the bulge below the Wash, and the tenth and last follows the coast of France across the English Channel. Another set of N40W directionals split Ireland into three parts, box in the Hebrides, and cut the Scottish mainland off from the Orkneys. The coast of Scotland below Moray Firth is well defined by a W2N that also splits apart the Hebrides to the west. The Scottish east coast below the firth is bounded by a N7E that can be extended north through the Shetland Islands, and south to cut off Wales from England. A number of other coastline features in the area may be traced with other directionals from the network.

These British Isles were not separated from the European continent until the end of the last ice age about 10,000 years ago when the continental ice sheets melted and the oceans rose to cover low-lying land areas. For the next few thousand years the people in the British Isles roamed the country hunting and fishing with no permanent settlements or farming communities. They had no pottery, but did develop the art of chipping flint rock to make tools such as axes, scrapers, knives, arrowheads, and awls. Then about 6,500 years ago there was a large migratory movement of people westward across Europe and into the British Isles who carried with them a culture that included farming, pottery, and probably the Indo-European dialects that were the roots for English and most of the other European languages. This is one of many migrations that have occurred in the past, and there is no agreement as to why large groups of people should suddenly leave home to wander across the world. One possibility suggested is a large volcano eruption that could change weather drastically as the volcanic dust spread out to block out sunlight and lower temperatures, causing crop failures over large areas. It is interesting that a large eruption did occur about 6,500 years ago when the Indo-European movement took place. What is now Crater Lake in Oregon was formed at that time when Mount Mazama exploded and blew off a large por-

tion of its summit. Perhaps this caused crop failures that forced the Indo-European farmers to pack up and move out, looking for a better place to farm.

Soon after the arrival of the new people in the British Isles settlements began to appear, and a little later they began to build the large earth and stone tombs and monuments that are scattered over the country. During a period lasting more than a thousand years and ending around 3,500 years ago, hundreds of large circles, ovals, and lines were laid out with mounds of earth or free standing stones, some with sight-lines designed to mark the extreme positions of the Sun and Moon on the horizon. As this period of monument construction came to an close, another migration moved west across Europe, bringing with it bronze tools and the Celtic language, the root of Irish, Scottish Gaelic, and Welsh that are still spoken in some parts of the British Isles. For a time the Celtic culture dominated most of Europe, until it was squeezed out by Rome moving up from the Mediterranean and the Germanic Tribes moving down from the Baltic. In the end, it was only in Ireland in the far west on the edge of the Atlantic that the Celtic culture was preserved in a relatively pure form, and much of that culture still colors the folklore of those living today among the beautiful green hills of Erin.

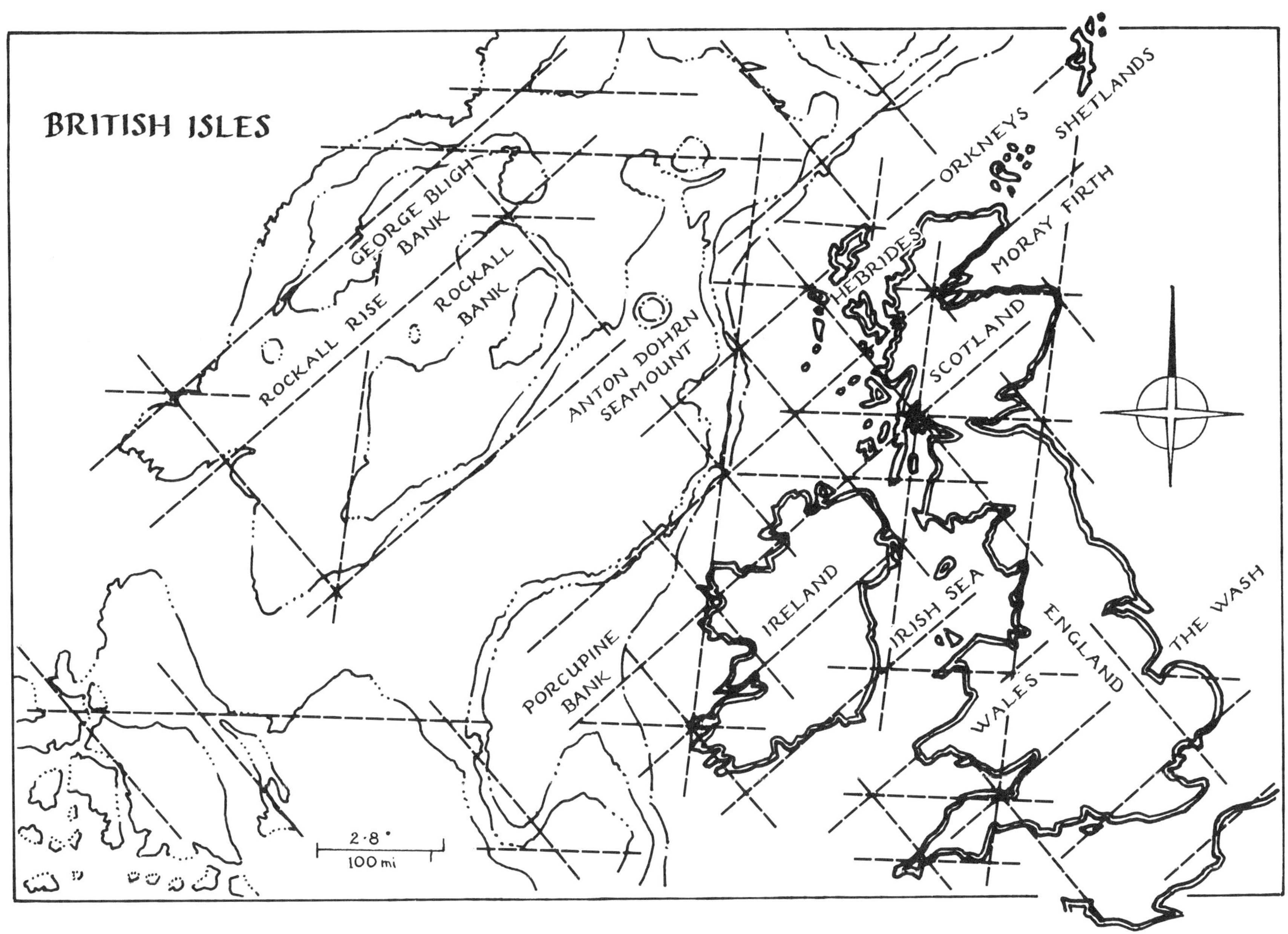
BRITISH ISLES
GEORGE BLIGH BANK
ROCKALL RISE
ROCKALL BANK
ANTON DOHRN SEAMOUNT
HEBRIDES
ORKNEYS
SHETLANDS
MORAY FIRTH
SCOTLAND
IRELAND
IRISH SEA
ENGLAND
WALES
THE WASH
PORCUPINE BANK
2·8°
100 mi

4-3 Indian Ocean Floor

This shows the upper part of the Indian Ocean from the horn of Africa, just off the edge of the figure on the west, to the tip of Sumatera on the east, and from the southern tip of India in the north to Mauritius in the south. The single lines in the figure represent subsea contours, and the double lines outline the few land areas shown. The left half of the figure is dominated by the Carlsberg and Indian Ocean Ridges which are long, linear mountain ranges lying on the ocean floor. The confused smear of contours in the upper left defines the Owen Fracture Zone, marking the beginning of the Carlsberg Ridge that strikes off towards the southeast between two N40W directionals separated by a 2.81 deg interval. In this section of the Carlsberg Ridge the contours are remarkably straight and fit the directional orientations and hierarchical interval almost exactly. The smooth, straight contours run for about 500 miles where the ridge appears to suddenly shift about 2.81 deg to the northeast. Toothed contours appear at the shift representing N50E transverse faults crossing the ridge. A short distance further on the ridge suddenly changes direction from N40W to N7E, but with the transverse faults maintaining the previous N50E orientation. The Carlsberg Ridge then continues south along N7E for a few hundred miles until it ends where it joins the Indian Ocean Ridge that angles back towards the lower left along N50E. The Indian Ocean Ridge can be further extended southwest off the figure, following N50E to a point well below the southern tip of Africa where it meets the Mid-Atlantic Ridge which cuts apart the Atlantic as it zigzags north all the way to Greenland.

The Indian Ocean Ridge is also called the Southwest Indian Ocean Ridge because there is a third ridge in the Indian Ocean that strikes off from the juncture of the Carlsberg and Southwest Ridges along N40W towards the lower right. The contours for this third ridge do not show up because it is relatively broad and flat with little relief. The place where the three ridges meet near the center of the figure is called a triple point. The Carlsberg Ridge branches up N7E from the point, the Southwest Ridge down left N50E, and the Southeast Ridge down right N40W. In the lower left of the figure are the Mascarene Plateau and the Seychelles that jut out from the Indian Ocean Ridge towards the northwest along N40W, paralleling the upper Carlsberg Ridge. The Mascarene Plateau is called a microcontinent because it is made up of thick continental crust in contrast to the surrounding thin ocean crust, and it has little in the way of volcanic or earthquake activity. Small microcontinents of this kind, that appear to be broken off bits of continents, are found in other parts of the world and include the Rockall Rise in the Atlantic that was shown in the previous figure.

Just east of the Carlsberg Ridge are the Maldive Islands on the Chagos-Laccadive Plateau, which show some of the characteristics of the Mascarene Plateau and may also be a microcontinent. The trend of the Maldive contours is initially N7E just north of the Indian Ocean Ridge, but it gradually swings off toward the west as it runs north along the northwest Indian coast that trends N40W. East of the Maldives is a closed contour that defines a long, almost perfectly linear subsea mountain range trending N7E which is called the Ninetyeast Ridge. It is the longest such feature on Earth, stretching for more than 3,000 miles with a width of about 125 miles, and rising up 6,000 to 10,000 feet above the ocean floor. In contrast to other ocean ridges, it shows no sign of volcanic or earthquake activity, and there are no known transverse faults that cross the ridge. It was not discovered until 1962 because of its depth of more than 10,000 feet below sea level, but only ten year later the Glomar Challenger was able to bring up five rock samples from bore-holes drilled along the crest of the ridge. The Glomar Challenger was designed specifically for this type of work, and could drill into the sea floor in water depths of up to four miles while it held its position directly over a location for many days in rough seas. This required computer controls to continually adjust a number of propellers that could move the ship in any direction. A sonar beacon that continually sent out sound pulses was let down on to the ocean floor next to the bore-hole. As the pulses were received by the ship, they were automatically sent to the computers which calculated the changes required in the positioning propellers to maintain the ship over the location, and which sent out the signals to implement the action. It was then possible to let down drill pipe and drill out a cylindrical chunk of material called a core up to 30 feet long from the ocean floor, then bring it back up to the ship where it could be analysed. The cores obtained from the Ninetyeast Ridge contained coal, peat, and fossils of shallow water animals, indicating that the underwater mountain peaks probably rose above sea level at one time as a string of islands.

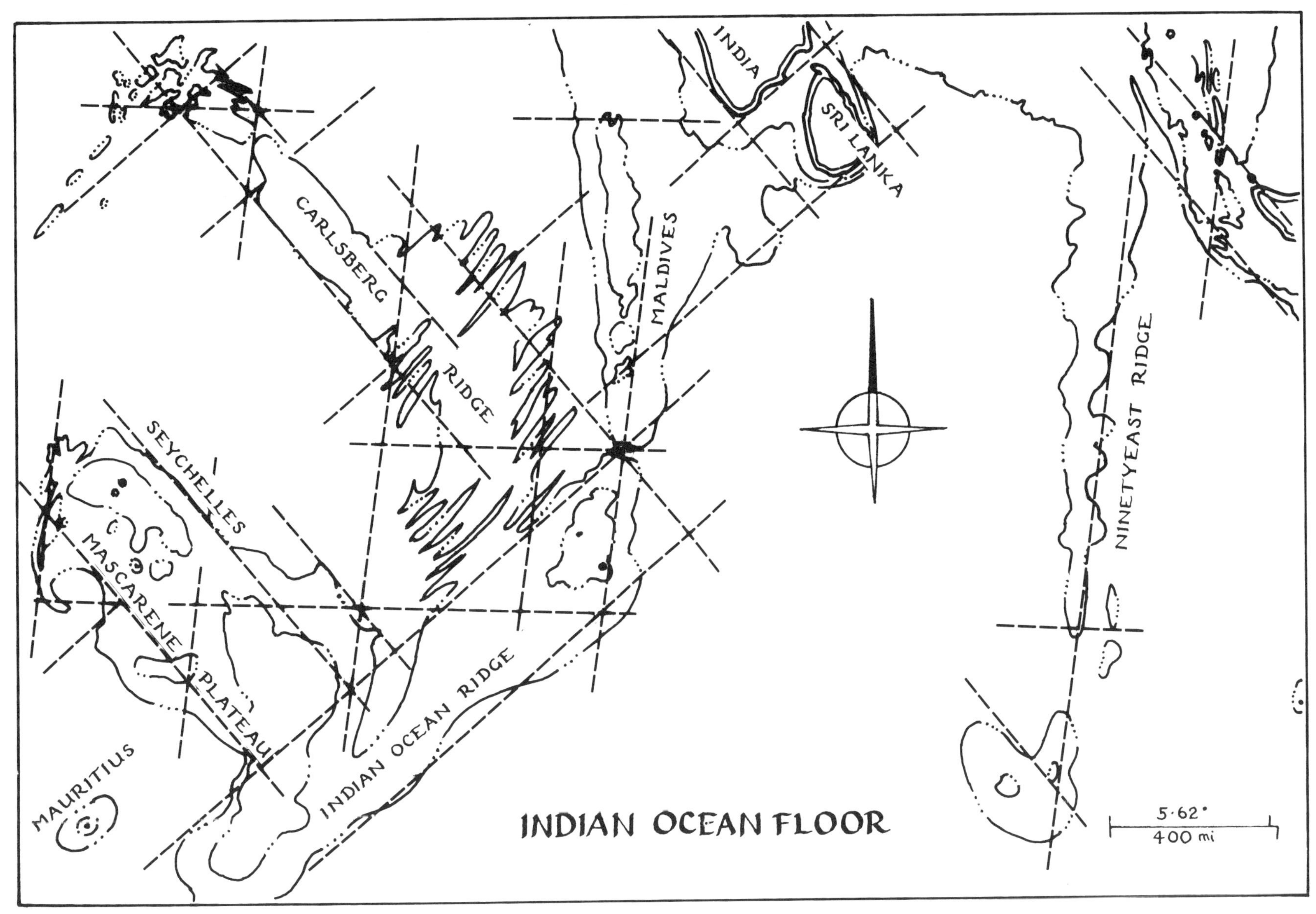
INDIA
SRI LANKA
CARLSBERG
RIDGE
MALDIVES
SEYCHELLES
MASCARENE
PLATEAU
MAURITIUS
INDIAN OCEAN RIDGE
NINETYEAST RIDGE
INDIAN OCEAN FLOOR
5·62°
400 mi

4-4 California Fault System

This shows the pattern of the extensive fault system that covers all of California. The area is just east of the Pacific Ocean and is part of the volcanic and earthquake prone Ring of Fire. An average of 500 earthquakes that can be felt and one of destructive magnitude occur each year in California. Although earthquakes may happen almost anywhere in the state, most of the large, destructive quakes have been centered on the longer, more well defined fault zones that are shown as heavier lines. These zones are made up of thousands of smaller faults oriented roughly in the same direction and clustered together within a narrow path that may run for miles.

The well known San Andreas Fault Zone on the west runs for more than 600 miles approximately N40W along the Pacific coast. It probably starts near Eureka in the north, cuts down through the San Francisco Bay area, then passes east of Santa Barbara and Los Angeles, and on south to the Gulf of California in Mexico. The great Tejon Pass quake of 1857 was centered on the San Andreas Fault near its intersection with the Garlock Fault east of Santa Barbara, and the famous San Francisco earthquake of 1906 was also caused by movement along the San Andreas. Just east of San Francisco Bay are the Hayward and Calaveras Faults that were responsible for the 1836 and 1868 earthquakes, two of the largest to hit northern California. Further south near Santa Barbara are the Santa Ynez group of faults that have been the center of a number of quakes of various sizes. Northeast of Santa Barbara is the Garlock Fault that has not produced any large earthquakes over recent history, even though it is the second longest in California. However, just north of the Garlock is the small White Wolf Fault that was also considered dead until the major Arvin-Tehachapi earthquake hit in 1952. Further east is the Sierra Nevada Fault that parallels the mountain range of the same name and was responsible for the 1872 Owen's Valley earthquake, the largest ever to hit

California in historic times. A number of large earthquakes have been centered east of Los Angeles on the San Jacinto Fault, which may be the main southern branch of the San Andreas Fault Zone.

The size of an earthquake is measured by both magnitude and intensity. The magnitude of an earthquake represents the relative size of the source underground where the rocks first shift suddenly and cause the disturbance, and the intensity represents the amount of destruction that could occur at various places on the surface. The magnitude is measured on the Richter Scale that can run from three which is hardly felt, to eight which may cause total destruction. An increase of one on the Richter Scale represents a ten-fold increase in quake size and a sixty-fold increase in energy released, so that moving from a magnitude of four to eight on the scale means an increase of ten million times in released energy. The intensity or destructive capability on the surface is measured on the Modified Mercalli Scale and is highest just above the earthquake source where destruction is greatest, and is lower for locations further away. The Mercalli Scale runs from I where there is no damage, to XII where there is total damage. The three large California earthquakes of Tejon Pass in 1857, Owen's Valley in 1872, and San Francisco in 1906 were all of intensity X or greater in surface locations nearest the source. It was the wide destruction by earthquake and fire in the San Francisco quake that focused attention and initiated a concentrated study of earthquake behavior. The effort provided the basis for the development of building standards in earthquake areas and marked the beginning of modern seismology. However, as yet there has been little success in forecasting the timing and magnitude of earthquakes, and they remain unpredictable along with much of nature.

Most of the California fault system is contained in a rectangle open to the northwest. The two long sides of the rectangle run N40W following the east and west boundaries of the state, and the base is closed off by the White Wolf Fault that follows a N50E directional. The southwest side of the rectangle is well

defined by the San Andreas that closely follows a N40W directional from San Francisco south to the White Wolf, establishing a linear grain for the smaller faults just west. The other side of the rectangle follows another N40W directional lying 2.81 deg east of the San Andreas and running just below and parallel to the Furnace Creek Fault. It forms the eastern border for a complex of faults located across the rectangle from San Francisco. Southeast of the Garlock Fault the pattern seems more complex with a profusion of many relatively small faults. However, a number of directionals cut through the area and define the boundaries of several smaller fault systems. Along these directionals the faults tend to begin, end, branch, and change direction, consistent with the fault system characteristics described in previous chapters.

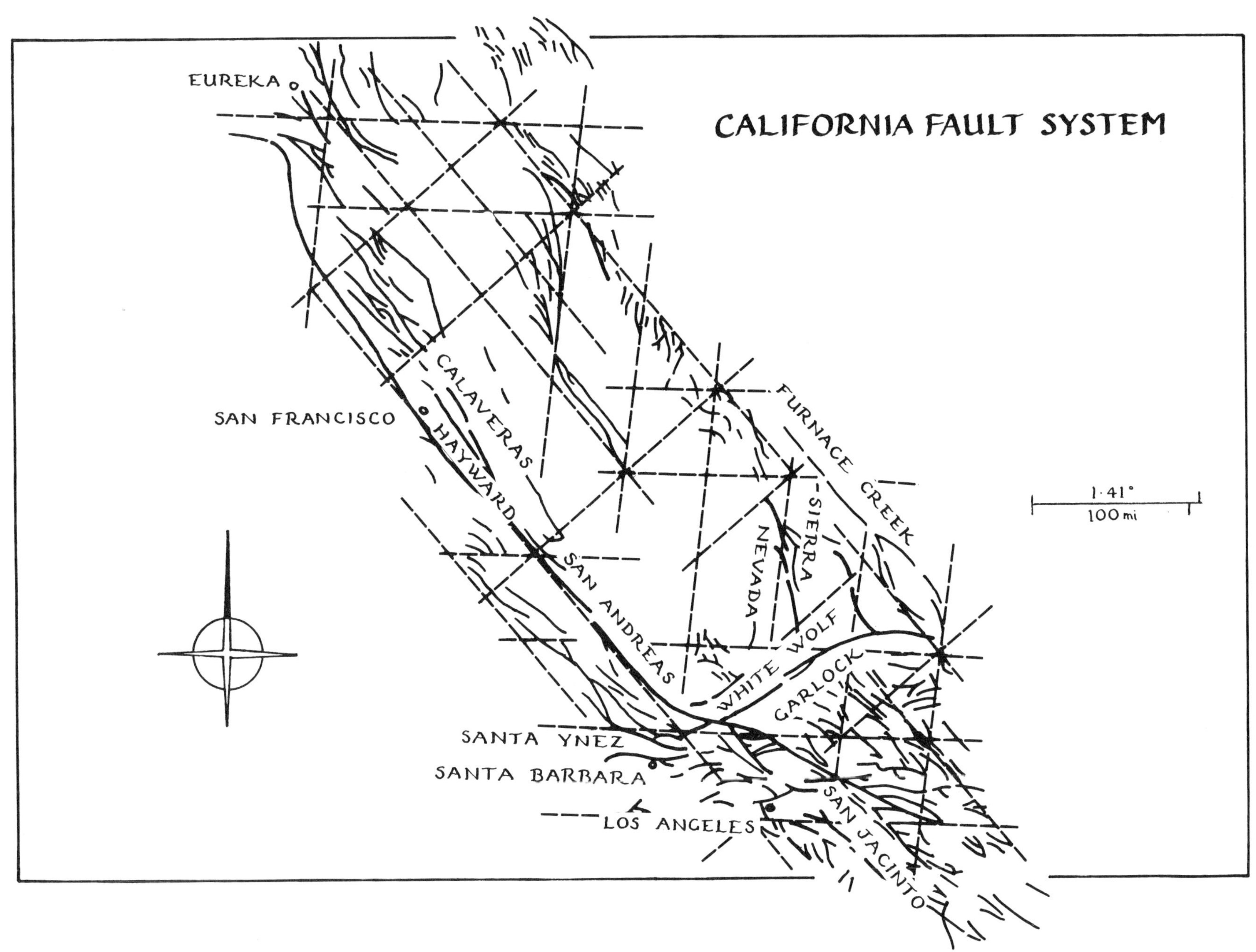
CALIFORNIA FAULT SYSTEM
EUREKA
SAN FRANCISCO
CALAVERAS
HAYWARD
SAN ANDREAS
FURNACE CREEK
SIERRA NEVADA
WHITE WOLF
GARLOCK
SANTA YNEZ
SANTA BARBARA
LOS ANGELES
SAN JACINTO
1·41°
100 mi

4-5 *Zagros Fault System*

This shows the Zagros Fault system that overlays the Zagros Mountain Range in Iran. The range covers an area over 550 miles long and 150 miles wide, running parallel to the northeast coast of the Persian Gulf. The individual mountains are generally long and narrow with steep flanks that rise to hights of 12,000 feet or more. Streams flow between the mountains following the grain of the land, but often change course abruptly and cut directly across an adjacent ridge, creating many narrow gorges running perpendicular to the overall grain. Southwest of the Zagros is a low dry plain that grades into salt flats as the coast of the gulf is approached. The area has a Mediterranean climate with the rain coming almost entirely in the winter months, often bringing heavy spring flooding to the low coastal plain. Summer in the lowlands is dry and very hot with temperatures often above 120 degrees Fahrenheit that bake the plain to a dusty brown. But with the coming of the winter rains, the parched land is turned into a cool green carpet of grass dotted with colorful flowers. The black tents of the Bakhtyari nomads appear at this time as they move their flocks down from the summer ranges in the mountains to graze on the green winter plains. The cities of Shiraz to the east and Esfahan to the north are in the mountains, with Ahwaz and Abadan situated on the low coastal plain. Ahwaz is the center of the oil producing industry, with many giant oil fields and highly productive wells located close by. Abadan on the Persian Gulf has an oil refinery that was once the largest in the world, but much of the facility is old and out-of-date, and much has been destroyed since September, 1980 when Iraq invaded Iran and war began.

This war is not the first between the people of the high Zagros land on the east and the Mesopotamia people from the Tigris and Euphrates River valleys on the west. There has been repeated conflict between the two since the beginning of written history, and probably before. Around 3,000 B.C. writing was discovered by the Sumerians in Iraq and quickly spread to their neighbors the Iranian Elamites, whose capital city Susa was located just west of Ahwaz. Much of the first writing told of conflicts and conquests, with the Elamites overrun several times by various powers from Iraq, including Ur in 2,700 B.C., Sargon the Great in 2,300 B.C., Hammurabi of Babylon in 1,750 B.C., Assyria in 1,200 B.C., Nebuchadrezzar of Babylon in 1,100 B.C., and Ashurbanipal of Assyria in 650 B.C., who completely destroyed the Elam capitol of Susa. During this same period there had been a gradual movement of Indo-European speaking people called Aryans into the north of Iran. They migrated south to establish the Mede and Persian Empires, and to eventually clash with the Assyrians from Iraq around 700 B.C., about 50 years before the Persian Cyrus the Great expanded the empire to include Babylon and Egypt, and Darius the Great added Afganistan and Pakistan on the east. This large Persian Empire lasted almost 200 years before it was conquered by Alexander the Macedonian from Greece. Much of this early history came from scattered clay tablets dug up from various places and may represent only a fraction of the conflicts that occurred. Certainly, the smaller raids and border disputes were never recorded on tablets, and many records must have been destroyed as cities changed hands from time to time. The war between Iraq and Iran might well be viewed as a continuation of a 5,000 year conflict. In the light of history, some things change very little.

Consistent with previous patterns, few of the Zagros faults trend along directionals, but rather appear to be contained within the directional network. The Main Zagros Fault can be traced for several hundred miles across the page from the upper left down to the lower right. Just above Shiraz it divides into a long branch above and a short branch below. A N50E directional crossing the branch point can be extended northeast to the intersection of a branch point on another major fault, and to the southwest where it cuts across an offset in an "S" shaped fault that lies on a N7E directional just west of Shiraz. This N7E directional with it's adjacent fault marks the eastern boundary of the area containing the largest and most productive oil fields in Iran. In the extreme lower right the Main Zagros Fault ends against a N50E directional that extends south to cut off several curved faults. Just above these curved faults is a N40W directional that can be extended to the northwest, crossing the Main Zagros Fault at it's lower end, then passing just north of Esfahan and continuing on to form the lower boundary of a set of faults in the upper middle of the page. Many other directionals may be observed that bracket the ends of individual faults. Another interesting observation concerns the orientation of the Main Zagros Fault which runs north 50 deg west, just 10 deg off the N40W diagonal directional. This makes the fault the mirror image of the N40W directional which could be more than an odd coincidence. This observation will be brought up in a later chapter and the possibilities discussed

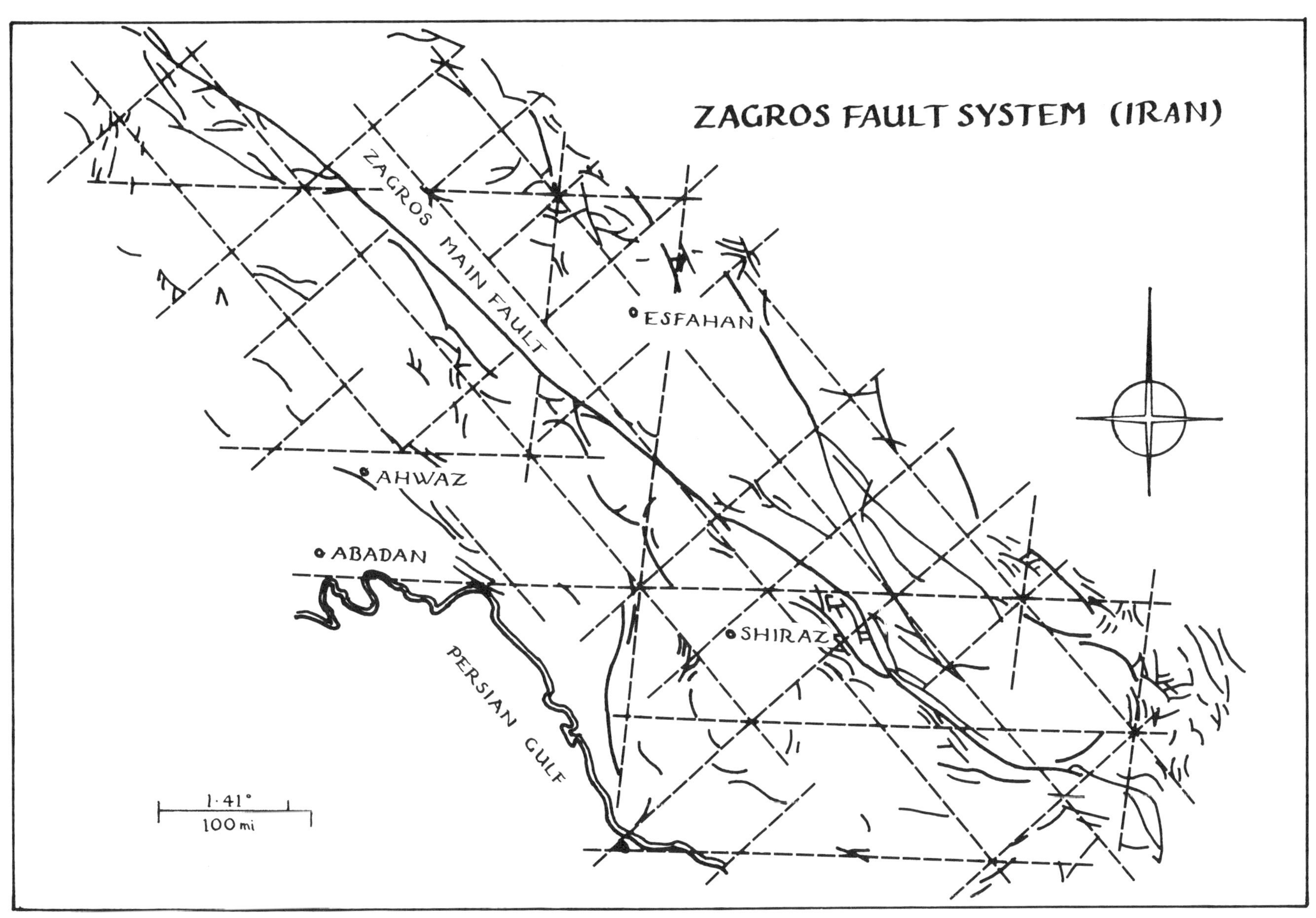

ZAGROS FAULT SYSTEM (IRAN)
ZAGROS MAIN FAULT
ESFAHAN
AHWAZ
ABADAN
SHIRAZ
PERSIAN GULF
1.41°
100 mi

4-6 *Laki Lava Flow*

The lava flows from the 1783 eruption of Laki Volcano in Iceland are shown here. The illustration was traced from Figure 61 on Page 111 of "Volcanoes" by Robert and Barbara Decker, 1981. About five cubic miles of lava was produced over eight months, the largest lava flow in historic times. The flow poured out of a 20 mile long, N50E fracture zone located along the upper left boundary of the flow. It was concentrated in two tongues which poured out of the ends of the fracture zone. These three features form a box open to the southeast with the sides formed of the two N40W trending lava tongues on the northeast and southwest, and the base on the northwest following the N50E fracture zone. The lava flow that forms the lower side of the box tended to puddle when it reached the foot of the volcano and spread out on the plain below, forming a cross within N40W and N50E directionals. Both tongues also follow N7E directionals in the upper parts where they first issued from the fracture zone. The flow characteristics are much the same as found in Hawaii, with directionals providing a preferred orientation pattern within which the lava flow course shifts rather abruptly from one directional to another.

It was the 1783 Laki eruption that provided the conditions leading to the first suggestion that a large eruption might effect weather. Ash and dust from the prolonged activity of Laki settled over Iceland and North Europe in a blue haze of dry fog. In Iceland the hay crop was destroyed, probably from poisonous fluoride in the ash, and with the loss of food over half the cattle, sheep, and ponies died. As a result of the enormous loss of livestock and exceptionally cold winters in 1783 and 1784, about one-fifth of the population died of starvation. It was the always curious Benjamin Franklin of the American Revolution, living in England at the time, who suggested that the colder winters might be related to the Laki volcanic activity. He reasoned that the dusty blue haze in the air could reduce the amount of sunlight reaching Earth, resulting in a temperature drop. Although scientists agree that volcanic dust would cause some slight cooling, at least in the near vicinity of a large eruption, there is still no consensus on whether the temperature drop is at all significant in relation to the many other complicated mechanisms that effect the weather.

Iceland is an up-raised part of the Mid-Atlantic Ridge where volcanic activity is concentrated. In Icelandic folklore there is a fire giant called Surtur who guards the flame world Muspell in the south, and his name was given to the new island of Surtsy, created by an eruption that began in 1964. The volcano Hekla, known as the Gate to Hell in Medieval Europe, has erupted at least 20 times since A.D.900 when Iceland was settled by the Norsemen. The year 1875 was another time of famine for Iceland when the hay crop was poisoned from dust blown out of the volcano Askja. Some volcanoes in Iceland, such as Katla, are covered with permanent ice caps that can block the release of hot ash during an eruption and cause the build-up of large quantities of hot melt water and steam under the thick glacial ice. When the ice cover is finally breached, huge quantities of trapped water are suddenly released in a destructive flood called a Jokulhlaup in Iceland, and a Lahar in other parts of the world. A huge volcanic mud flow of this type occurred recently in Colombia on November 13, 1985. The Volcano Navada del Ruiz erupted below a glacial cap and melted about 10% of the ice before it broke through and flooded down the mountain. The Lahars poured down two rivers, wiping out several villages and burying some 25,000 people. It was the greatest volcanic disaster since May 8, 1902 when a cloud of hot ash from Pelee in the West Indies swept through St. Pierre and destroyed the entire city with its 28,000 people.

Krafla Volcano in northeast Iceland is the site of a recent geothermal development. Geothermal energy comes from steam or very hot water produced from wells drilled into hot fractured rock underground. Geothermal development can be a risky business due to the uncertainty of how much hot fluid the underground rocks might hold, and what the flow rate of each expensive well might be. Another unknown is just what type of hot fluid the rocks contain, and that can range from superheated steam to hot water. When the Krafla development occurred there was little information available, and the plant was designed for hot water. Unfortunately, after the money was spent and development complete, it was found that the situation was considerably more complicated, with the wells producing both steam and hot water. Also unfortunate was the unpredictable resumption of volcanic activity after a quiet period of some 250 years. All this contributed to the completion of a 60 megawatt geothermal plant in 1977 that could produce no more than 7 megawatts. Geologic knowledge of the Earth is very limited and generally broad in scope, and there will always be uncertainty and relatively high risks for those developments dependent on below surface estimates. Perhaps it is as much the degree of uncertainty and luck as the amount of sophisticated engineering and planning that determines much of the success or failure of geologic based projects. But that is the nature of the game.

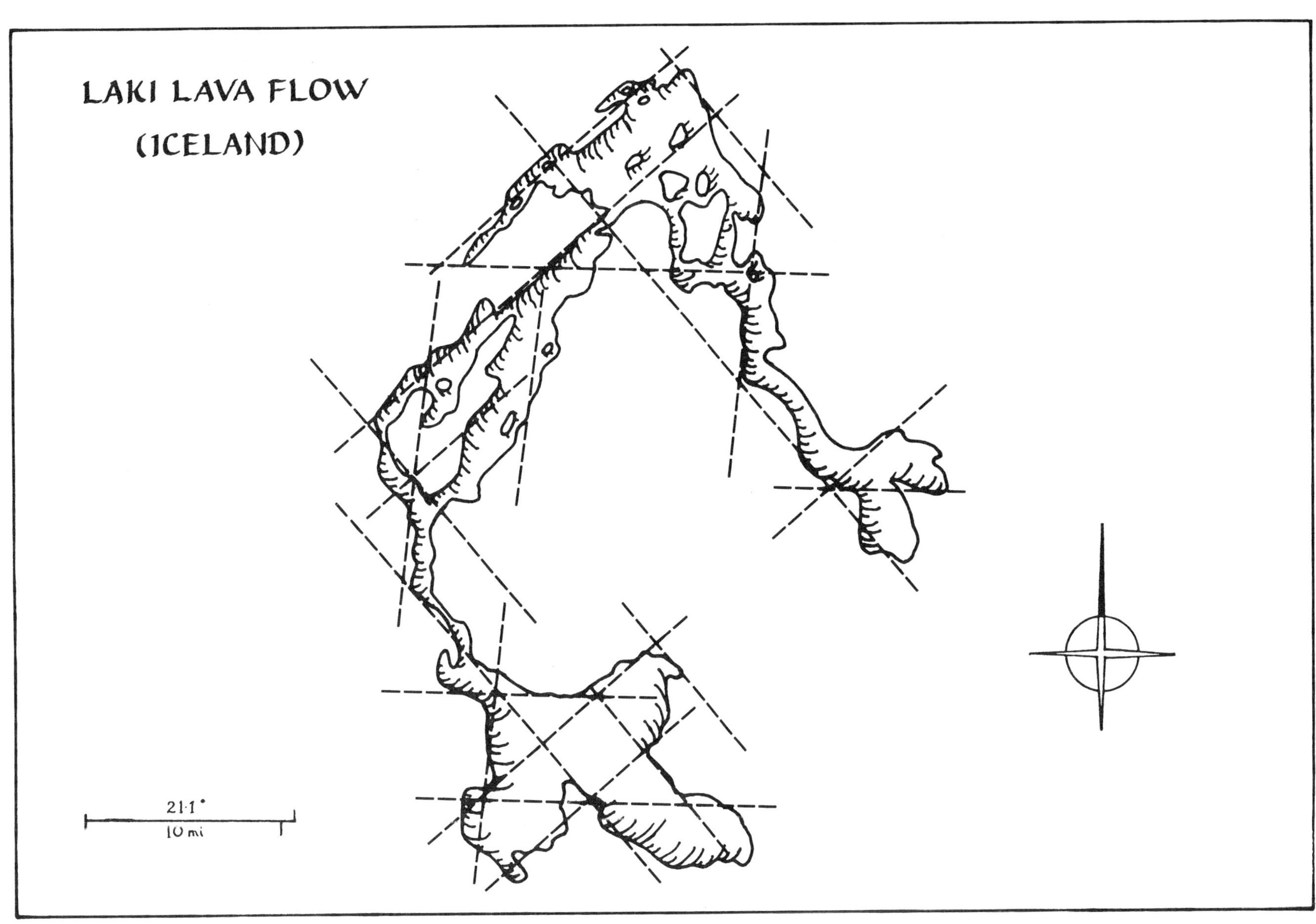

LAKI LAVA FLOW
(ICELAND)
21.1°
10 mi

4-7 Mount St. Helens Pyroclastic Flow

This shows the pyroclastic flows from the 1980 eruptions of Mount St. Helens which is located in Washington State. It was traced from Figure 295, Page 503 in "The 1980 Eruption of Mount St. Helens," United States Geologic Survey, Geological Survey Professional Paper 1205, 1981. The authors were Peter D. Rowley, Mel A. Kuntz, and Norman S. Macleod. The volcano is one of several in the Cascade Range that lies in the northwestern United States and borders the Pacific Ocean. It is part of the Ring of Fire circling that Ocean, and runs north from Lassen Peak and Mount Shasta in California to Crater Lake and Mount Hood in Oregon, and on to Mount St. Helens, Mount Rainer, and Mount Baker in Washington. Several of the Cascade volcanoes have been active in recent times, with Mount Baker erupting in 1870 and Mount Rainer in 1850. Before the 1980 eruption, Mount St. Helens last activity was in 1857, but geologic studies showed that it had erupted at least 20 times over the last few thousand years and was considered the most active volcano in the continental United States. In 1978 a study by the U. S. Geological Survey forecast that Mount St. Helens would probably erupt within the next 100 years, and possibly before the end of the century. The 1980 volcanic activity was not all that unexpected.

Prior to a volcanic eruption there is generally a flow of hot gas and magma from deep in the earth up into shallow reservoirs under the volcano, and the flow path is reflected in small magnitude earthquake swarms as rocks are moved aside or fractured. Earthquake swarms began under Mount St. Helens in March 1980, and the number gradually increased, indicating the movement of lava below the volcano. Within a few days a summit crater began to develop along with small explosions of steam from subsurface water that had been heated until it flashed into steam and forced its way to the surface. A bulge also began to develop on the north flank just below the surface. The bulge continued to grow larger until

May 18, 1980 when the volcano violently erupted.

The eruption was not only seen by nearby observes, but was also photographed. A small earthquake started a huge avalanche on the north flank bulge that had been stressed to the point of break-up. About half a cubic mile of rock was torn out of the north flank and plunged down the mountain into the valleys below. As the bulge was ripped apart, high pressure steam and volcanic gasses were blown out laterally in a huge violent explosion. The steam, gas, and avalanche material were churned up into a liquefied mass of mud that flowed down the mountain at a speed of close to 200 miles per hour, and the mud flow was followed by a dense hot cloud of steam mixed with volcanic material that blasted out to the north and destroyed more than 200 square miles of forest. The eruption tore about 1,200 feet off the top of the mountain and dug out a crater over 2,000 feet deep. Not long after the explosion, pyroclastics erupted and flowed north over the devastated area.

Pyroclastics are solid fragments of volcanic material that vary from fine dust to rocks of several tons. The pyroclastics may either be blown up into the air or flow out of the volcano and down the mountain as a hot liquid mass. A glowing flow of pyroclastic hot gas and ash is called a Nuee Ardente, and can travel up to 70 miles per hour destroying everything in its path. The hot particles carried by the gas are charged electronically and repel each other, so that they move along in the gas with very little frictional resistance due to the lack of contact between the particles. It was just such a hot ash cloud that erupted from Mount Pelee in 1902 and flowed down the mountain as a Nuee Ardente to completely destroy the city of St. Pierre.

The pyroclastic flows from Mount St. Helens poured out from the newly formed crater and spread out as far as Spirit Lake, along with numerous explosions as water in the path of the hot ash flashed into steam. Over the five months following the major eruption there were repeated smaller eruptions that generated numerous pyroclastic flows and formed

the pattern shown in the figure. The overall shape of the pyroclastic deposits resembles a letter "Y" standing on the crater. The stem of the "Y" just above the crater is oriented N7E with a width of 39.6 sec, which is one-half of the hierarchical interval 1.32 min given on the scale. This orientation defines the direction of the lateral explosion that blew off the top of the mountain during the eruption. The stem of the "Y" follows N7E for about two miles from the crater up to the point where the wings of the "Y" abruptly branch off the stem along N40W to the left and N50E to the right. The main area covered by the left wing is rectangular with the long sides formed of N40W directionals separated by a 1.32 min hierarchical interval. The short sides of the rectangle follow N50E directionals across a 2.64 min interval. The right wing is formed of two adjacent rectangles, again with sides oriented N40W and N50E, and with the upper rectangle approximately 39.6 sec wide and the lower about 1.32 min. The correspondence of the flows to the directional pattern is consistent with the strain analogy. A torsional strain pattern within the crust of the Earth would be reflected both in the internal construction of the volcano and in the external surrounding features, and any material released during the series of eruptions might be expected to correspond in some way to that basic strain pattern.

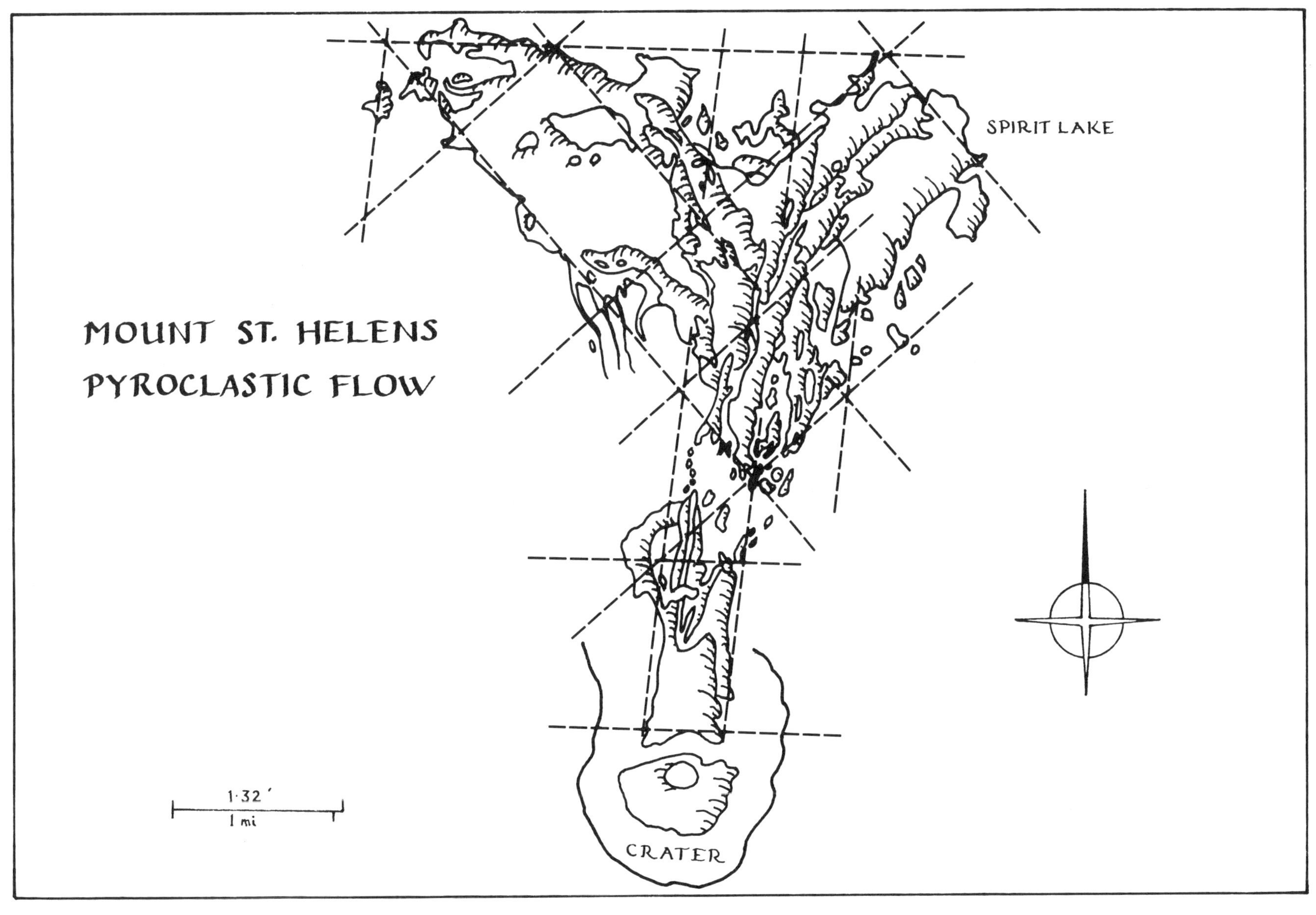
SPIRIT LAKE
MOUNT ST. HELENS
PYROCLASTIC FLOW
1·32´
1 mi
CRATER

4-8 Mississippi River System

This shows the Mississippi River basin stretching from Canada on the north to the Gulf of Mexico on the south, and from the Rocky Mountains on the west to the Appalachians on the east. It drains the largest area of productive farmland in the world, with a rich black loam soil spread over the central and eastern portions and watered by generous rainfall. The temperate climate with its temperature extremes may often be uncomfortable, but it is near perfect for agriculture. The cold winters kill off insects and weeds so that crops are given a new start each spring when the melting snow and early rains provide water for the new plants. During the hot summers the rains lessen and the sun pours out the energy needed for rapid growth. Then comes the cooler autumn which allows the fully grown crops to mature and ripen, ready for the harvest. This central area is known as the corn belt, and is often called the breadbasket of the United States. West of the cornbelt the rainfall diminishes, but is still enough to make it one of the largest wheat producing areas in the world. Perhaps the lucky inhabitants of the region are insufficiently aware of the immense fortune provided by Mother Nature, as compared to the scraps passed out to most others in the world.

Settlers began to move across the appalachians into the eastern Mississippi basin early in the 19th century after the Revolutionary War. However, the movement accelerated enormously during the "cold years" of 1812 to 1817 when crop failures were common over all of New England. The worst came during "the year without a summer" in 1816 when New England had frosts in all summer months. On June 7th there was a snowstorm with drifts over a foot high, and frosts occurred on the 8th, 9th, and 22nd of July. The summer ended with the September 27th "black frost," probably caused by the many forest fires that same summer. The drop in temperature may have been caused by the large amounts of dust thrown into the atmosphere by the eruption of the volcano Tomboro located east of Bali in Indonesia. Temperatures in both North America and Europe dropped more than five degrees below normal. The New England crop failures forced more and more people west across the mountains and into the greener pastures of the "Old Northwest" along the Ohio River. However, with all the green pastures there were unforeseen problems. With the spring floods came the stagnant water and clouds of mosquitoes and flies. Malaria was common, and cholera and smallpox spread quickly, sometimes wiping out entire settlements. But man is a tough, stubborn animal and the people continued to pour across the mountains into the fertile lands of the eastern Mississippi basin.

Movement west of the Mississippi began a little later. The immense area between the Mississippi and the Rocky Mountains was called the "Great American Desert," and for a time was left to the indians who depended on the large herds of buffalo that roamed the area. Then in 1803 came the Louisiana Purchase that transferred much of the area from France to the young United States, and the following year Louis and Clark started their journey up the Mississippi and over the Rocky Mountains to the mouth of the Columbia River on the western seacoast. They were followed by the "mountain men," those uncouth and rough fur trappers such as Jedediah Smith, Thomas Fitzpatric, Kit Carson, Jim Bridger, and the Sublettes. These were the men who found the mountain passes and made the trails that were later used by thousands of emigrants who followed their dream to Oregon and California. The Oregon Trail started near the branch of the Missouri and Platte Rivers and traced its way 2,000 miles across the great plains and the Rockies to the west coast. The flat plains looked like an endless ocean to the travelers, and they called their wagons "prairie scooners," and the long flat bluffs along the Platte River the "coast of Nebraska." The trip started in mid-May after the snow had melted and the land dried, and the Rockies had to be crossed before the winter snows. That required 15 miles each and every day whatever happened, with no waiting for breakdown, sickness, or death. In 1843 one thousand settlers traveled the trail, passing Chimney Rock, Scott's Bluff, and Ash Hollow along the Platte, then across South Pass, discovered by Jed Smith in 1823, and on to Oregon. A year later in 1844, the number of emigrants on the trail doubled and the following year the number tripled. The Oregon Trail has been truly called the greatest settlement road ever traveled within the United States.

The Mississippi River weaves back and forth across a central N7E directional dividing the basin into two regions with a N50E grain on the east and a N40W grain on the west. Just east of the Mississippi across an interval of 2.81 deg a N7E directional traces the lower reaches of the Wabash and Tennessee Rivers, with the overall N50E grain established further east along the upper part of the Tennessee and Ohio Rivers. This eastern N50E grain can be extended east to the appalachians and Atlantic seaboard. West of the Mississippi the rivers are longer and tend to branch off along N40W, which is the general grain within the Rocky Mountains and further west along the Pacific coast of North America. Several river features follow N40W directionals in the region. The system is bordered on the west by a N40W directional that defines the headwaters of a number of rivers, and further east the Arkansas, Missouri, and Upper Mississippi also follow N40W directionals. In addition, several N50E and W2N directionals can be seen along the river courses in the upper left of the page.

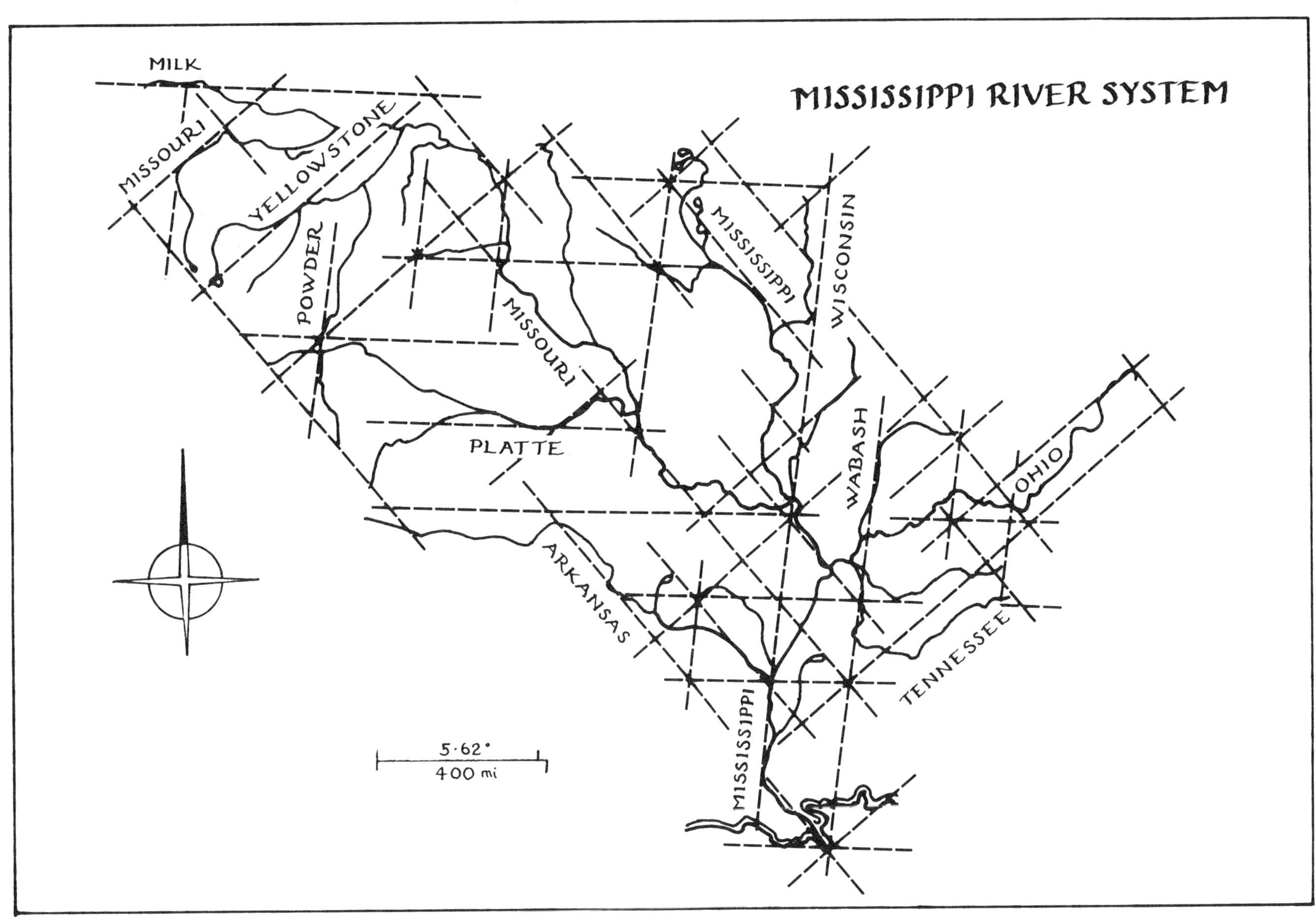
MILK
MISSISSIPPI RIVER SYSTEM
MISSOURI
YELLOWSTONE
POWDER
MISSOURI
MISSISSIPPI
WISCONSIN
PLATTE
WABASH
OHIO
ARKANSAS
TENNESSEE
MISSISSIPPI
5·62°
400 mi

4-9 Central Africa River Systems

This shows the pattern of the river systems in central Africa. The two major rivers in the area are the Congo, or Zaire, to the north, and the Zambesi to the south. Both river systems border on deserts, with the Sahara lying north of the Congo and the Kalahari Desert south of the Zambesi. The rivers flow in opposite directions with the Congo discharging into the Atlantic on the west and the Zambesi into the Indian Ocean on the east. Both rivers drain the lake complex in East Africa which lies within the Great Rift Valley, the largest feature of its kind in the world. The Rift Valley is formed of a strip of land some 30 to 40 miles wide that has sunk down between two parallel fault zones and runs for about 4,000 miles from the Middle East down to the Zambesi River. The lake system within the Rift Valley is one of the largest in the world, with Lakes Edward and Albert on the north flowing into the Nile, Lakes Tanganyika and Kivu at the center into the Congo, and Lake Nyasa on the south into the Zambesi.

The Congo River is about 3,000 miles long from the Atlantic to the East African Lakes and has many rapids, cataracts, and falls along its course, as do many of the African rivers. About 1,000 miles of the central Congo is navigable from Stanley Pool on the west where the Kasai River enters, to Stanley Falls on the east just beyond the Lomami River mouth. Further upstream the name of the river changes from Congo to Lualaba. The name Stanley that appears so often along the congo is from the great African explorer Henry Stanley who was born illegitimately in 1841 in Wales. He was ignored by his mother and spent his early life with relatives or in workhouses. At 15 he ran away and sailed as a cabin boy to the United States where he served in the American Civil War, and for a time as a seaman in the Federal Navy. He eventually became an international correspondent and gained a wide range of experience from his many travels. He also developed an independent and cantankerous disposition that made him uncooperative, intolerant, and hard headed, but also gave him the tenacity and singleness of purpose that carried him through the most difficult times. On his first trip down the Congo in 1876 he arrived back at the Atlantic after losing the other three in his party and half the African bearers he started with. But less than ten years later he finished a road to Stanley Pool, set up a chain of stations, and had steamers sailing on the 1,000 miles of navigable Congo. He made his first trip to Africa in 1871 on assignment with the New York Herald to find another great African explorer named David Livingston who had disappeared somewhere in Africa several years before.

In many ways, Livingston was the opposite of Stanley. He came from Wales, as did Stanley, but was born in 1813 at the beginning of a religious upsurge that spread over both Great Britain and America. He was caught up in the excitement and committed himself to go to China as a missionary doctor. However, the missionary work in China was virtually suspended during the Opium Wars of 1839 to 1842, and Livingston was forced to change his field of work from China to Africa. He sailed into Capetown in 1841 and spent the remaining 30 years of his life exploring the Zambesi basin and preaching his missionary message to the tribes of southern Africa. His wife died on the Zambesi during his 1858 to 1864 explorations, and he died nine years later on the shores of Lake Tanganyika. It was these two 19th century opposites from Wales, Stanley the fortune hunter and Livingston the missionary, who broke open the dark unknown of central Africa.

The Congo and Zambesi River systems are split apart by a dividing W2N directional that cuts cross the headwaters of the tributaries just below the center of the page. South of this dividing directional the Zambesi River follows a N40W directional for about 300 miles from its mouth on the Indian Ocean up towards the northwest. It then cuts sharply west along W2N for another 300 miles to the Kafue tributary which continues along W2N for another 200 miles. Here the Kafue cuts back abruptly to the northeast along N50E. From the mouth of the Kafue, the Zambesi changes course to N50E along Lake Kariba, and then curves up, first along N40W and then N7E. The Cubango tributary continues southwest along N50E, then breaks back abruptly to N40W, which is also the general orientation of the Cuito and Cuando tributaries in the same area.

North of the dividing W2N directional is the Congo River system which shows a N7E grain within its central portion with the upper Kasai, Sankuru, and Lomami all oriented along the grain. The Congo itself leaves the Atlantic generally following N50E towards the northeast to the mouth of the Oubangui which branches off to the north. A N7E directional can be followed from north to south that traces a portion of the Oubangui and an offset in the Congo. The N7E then passes along a lake on the Lukenie, and finally, much further south, to the Cubango headwaters. After leaving the Oubangui, the Congo curves around to the east and south, and then as the Lualaba descends in a stair step pattern, alternating between N7E and N40W. Finally as the Lualaba approaches its headwaters, it shifts abruptly to N50E parallel to a directional that also traces the upper reaches of the Kafue down to the southwest. These two rivers are also connected in the same area by a N7E directional which runs south along the Lualaba to the extreme end of the Kafue.

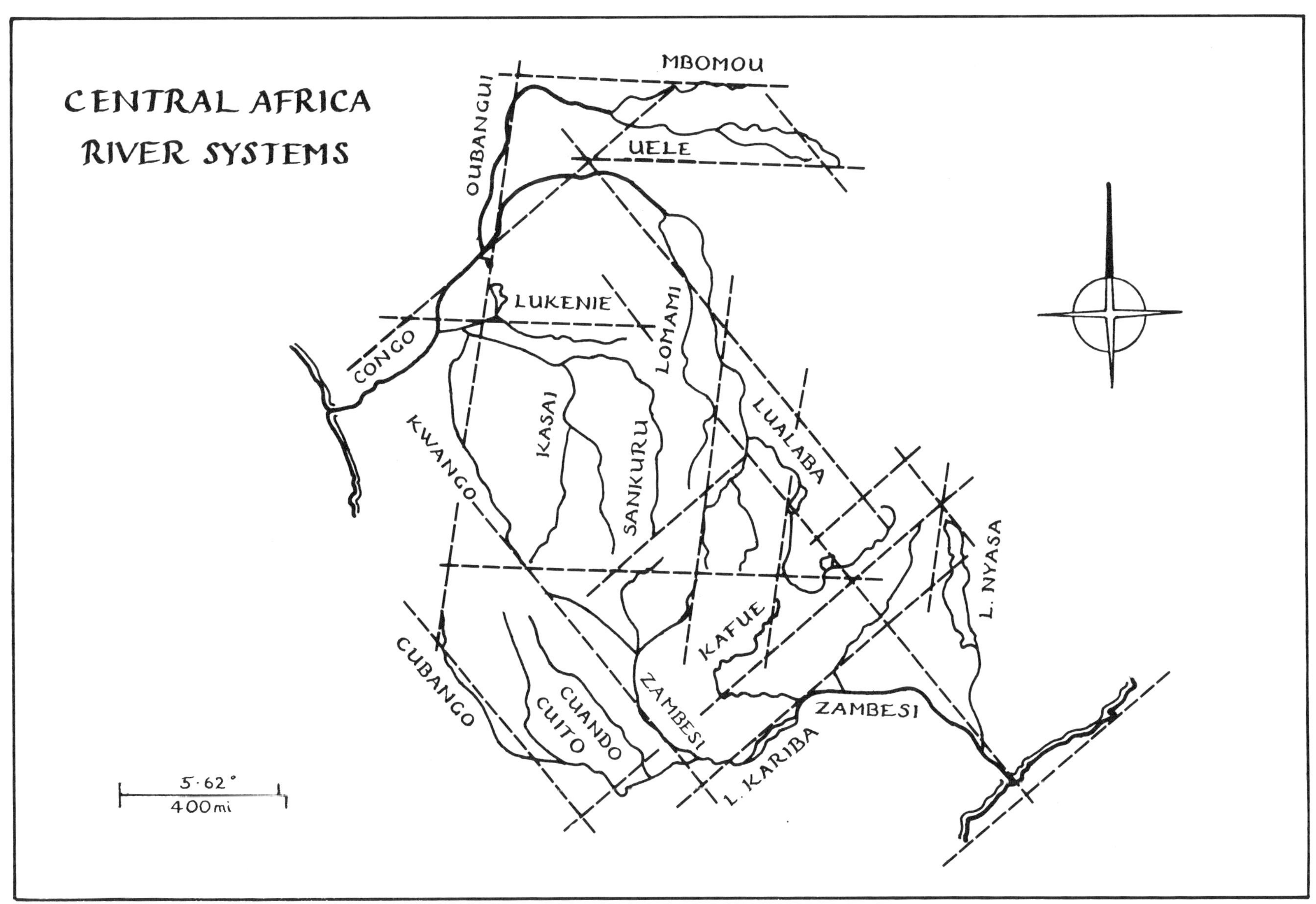

CENTRAL AFRICA
RIVER SYSTEMS
MBOMOU
OUBANGUI
UELE
CONGO
LUKENIE
LOMAMI
LUALABA
KASAI
KWANGO
SANKURU
L. NYASA
KAFUE
CUBANGO
CUANDO
CUITO
ZAMBESI
ZAMBESI
L. KARIBA
5.62°
400mi

4-10 Amazon River System

This shows the Amazon River basin, the greatest in the world by almost any comparison. The Amazon River is officially 4,000 miles long, and that makes it 150 miles shorter than the Nile which is the longest in the world. However, in terms of area the Amazon basin is about twice that of any other river basin and would cover about three-quarters of the continental United States. The basin holds about half of all rain forests in the world, and the water that drains from the basin into the Amazon is about one-fifth of the water runoff from the entire Earth. The Amazon carries the largest water flow in the world, with an average about twice that of the Congo and ten times that of the Mississippi, and in flood stage the average flow can double. The delta at its mouth on the Atlantic is about 250 miles across, and the water there is as pure as most tap water due to the settling of debris as the river widens and the rate of flow slows. Upstream from its mouth for the first few hundred miles the river is about 300 feet deep, allowing ocean going ships to travel 1,000 miles up to Manaus at the Negro River branch and river steamers to travel another 1,300 miles upriver to Iquitos, Peru at the Napo River branch. At Iquitos the name of the river changes from Amazon to Ucayali, and it continues upstream to its source high in the Andes near Cuzco, Peru, only 100 miles from the the Pacific Ocean.

The Amazon River system is bordered by clearly defined directionals with various localized grains displayed in the interior. On the north a W2N directional boundary runs across the entire continent, and on the west the system is bordered by an upper N50E and lower N40W that are separated by a cut-out, marking a N2W directional that follows the Amazon to the east. This western cut-out is also reflected in a parallel cut-out further west on the Pacific. The border on the south resembles a "W" formed of N50E and N40W directionals. The left half of the "W" encloses the Madeira tributary system and the right half the Tocantins' tributaries, with the center

apex in line with the upstream section of the Tapajos. The eastern border follows the general trend of the Tocantins along N7E. The delta at the Amazon mouth is enclosed in a directional triangle with a W2N forming the base, a N50E the northwest side, and a N40W the northeast side that cuts directly across the mouth of the river. Just west of the delta the Amazon drops slightly to the south, and then cuts down abruptly along a N50E directional that the Madeira River continues to follow over much of its course. From the Madeira outlet, the Amazon wanders west back and forth across a W2N directional all the way to the Napo where the Amazon changes to the Ucayali. The grain in the extreme northeast trends N40W from the Jari west to the Negro where the grain shifts to W2N further west. This grain shift is marked by the Branco at the top center of the page, which strikes north from the Negro along N7E, then turns abruptly west and follows W2N for 2.81 deg before it cuts back sharply to N7E. For most of the area south of the Amazon the grain is generally N50E, as defined by the lower reaches of the major tributaries. However, along the edges of the system the grain varies, following N40W on the west and south, and N7E on the east.

Several hierarchical intervals can be seen between adjacent directionals. The series of W2N directionals starting from the north border can be continued south at equivalent 2.81 deg intervals, first down to the triangle base below the Amazon mouth, next to the general trend of the upper Amazon, and further down to the sharp course changes in the Maranon, Ucayali, Jurua, Purus, Madeira, and Tapajos. The series continues south with the directionals marked by changes and terminations in the smaller tributaries. A series of N50E directionals across 2.81 deg intervals approximately follows the southern major tributaries starting with the Iriri on the east, and including the Tapajos, Madeira, Purus, and Jurua to the west. Just 2.81 deg to the right of the

N40W directional defining the southwest border of the river system is a parallel N40W that again follows tributary changes. Three N7E directionals separated by 2.81 deg can also be seen at about the center of the figure, with the left two directionals defined by the sharp course changes in the Branco, and the directional on the right by tributaries both north and south of the Amazon.

The correspondence between the directional pattern and three major river systems just described has many of the characteristics found in other geological systems. The directional grid acts as a framework constraining the orientations of the individual features within the system. The terminations and changes in both river courses and fractures tend to occur where the feature meets a directional, and the change is often quite abrupt. In addition, when a linear feature is offset it often shifts across a hierarchical grid interval to a parallel direction, reinforcing the grain in the general area.

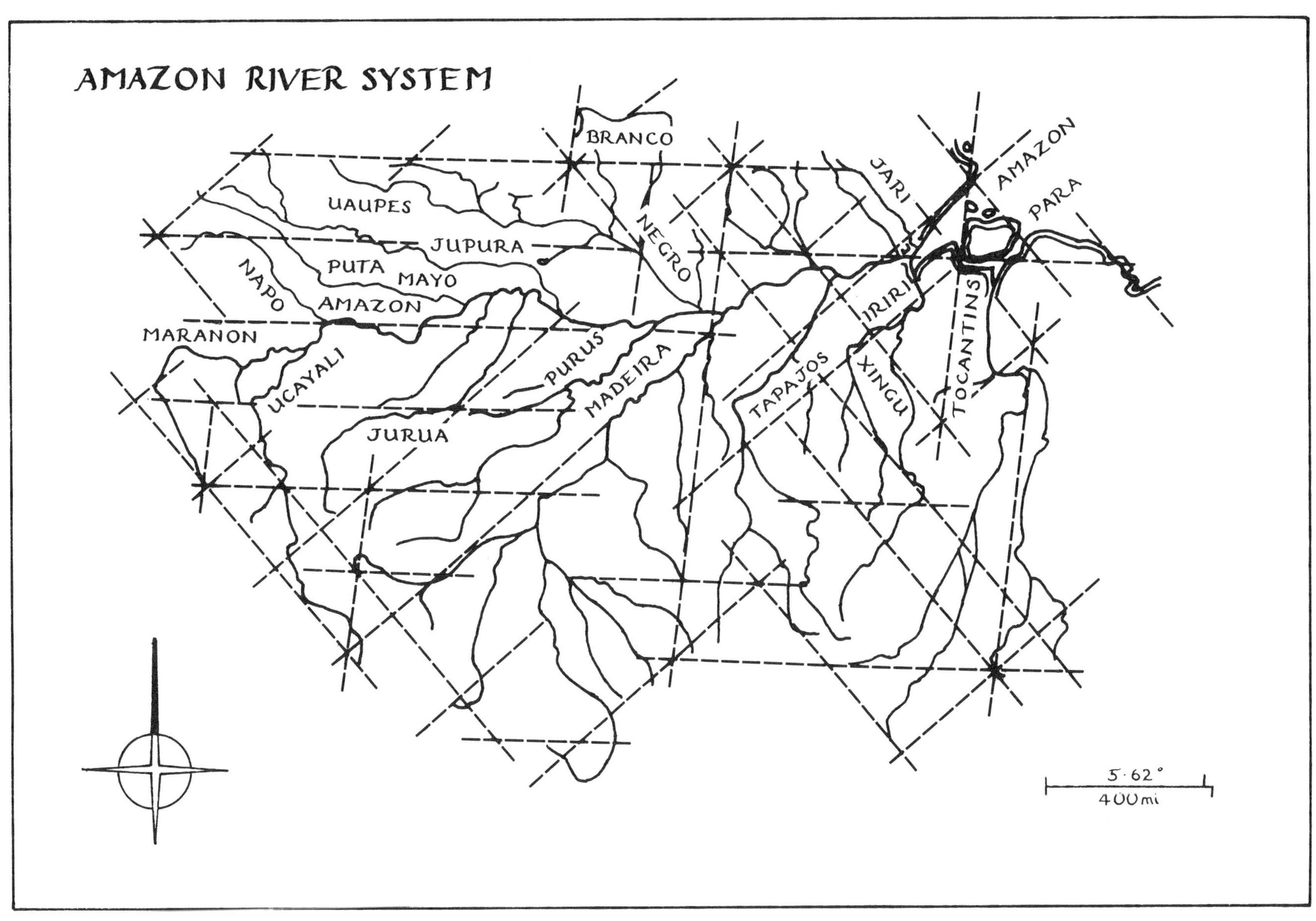
AMAZON RIVER SYSTEM
BRANCO
UAUPES
JUPURA
NEGRO
JARI
AMAZON
PARA
PUTA MAYO
NAPO
AMAZON
IRIRI
MARANON
UCAYALI
PURUS
MADEIRA
TAPAJOS
XINGU
TOCANTINS
JURUA
5·62°
400mi

In the first chapter a number of major coastlines were examined, and it was found that four orientations predominated. These four orientations were termed directionals and were measured as north 40 deg west, north 50 deg east, north 7 deg east, and west 2 deg north. These measured orientations were used to name the directionals N40W, N50E, N7E, and W2N, and this terminology has been retained throughout the book. All examples were traced from Mercator Projection maps in which true compass direction on the globe is represented by a straight line on the flat map. Other coastline examples were selected to illustrate combinations of these four directionals. It was also found that a coastline often shifted direction rather abruptly, either as a change to a different directional or as an offset with a continuation in the same direction. It was also found from other examples that a group of coastal features that followed a particular directional tended to be separated by equal intervals as measured in equivalent horizontal degrees on Mercator Projection. The equal intervals were found by measurement to approximate 2.8, 5.6, or 11.2 deg, which might represent a hierarchical series formed by doubling any member to obtain the next number in the series. Combining the empirical observations of all four directionals separated by equal intervals gave a repeating pattern of a rhombus and its diagonals, where a rhombus is defined as a figure with four equal sides that do not necessarily meet at right angles.

In the second chapter an analogy was presented to explain the repeating rhombus pattern. Mercator Projection is a projection from the axis of the globe directed out through the globe's surface, and on to the surface of a cylinder slipped over the globe and fitted around the equator. The cylinder with its projected features can then be split lengthwise and laid out to form a flat Mercator Projection map. A unit square grid was then traced on the Mercator map, and it was rolled back into a cylinder and slipped back over the globe. The cylinder was then twisted by rotating the ends in opposite directions, and again uncurled and laid out flat. It was found that the original unit square grid had been deformed into a repeating pattern of a rhombus and its diagonals, resembling the empirical pattern found on Earth. In mechanics the twisting force, or stress, is called pure shear, and the major deformation, or strain, is along the rhombus' diagonals, one of which is lengthened and the other shortened. The diagonals form right angles at their intersection point which theoretically represents a point of minimum strain that could provide the nucleus to initiate a hierarchical split of the unit rhombus into four similar figures. As the rhombus continues to split into finer and finer divisions, a series develops in which the rhombus' sides are repeatedly halved. Using this halving series and starting from the full 360 deg Earth circumference across the Mercator map, members 6, 7, and 8 of the series are 11.2, 5.6, and 2.8 deg, the same as the equal intervals found empirically between adjacent coastline features. It was postulated that the lines of extreme deformation in the rhombus strain pattern might represent paths of preferential flow that could either clog up or clean out as liquid passed through. Flow courses would then tend to shift from one strain directional to another as restrictions were encountered. If the crust breaks under deformation and faults are formed, they would generally follow imperfections rather than strain lines, but would tend to begin, change direction, and end along the strain lines.

The analogy and derived postulates were tested within the Hawaiian Islands and surrounding Pacific waters. All fault systems studied were found consistent with the postulates, from the huge trench system across the entire Pacific to the small dike system on Nihoa. Lava flows on the volcanically active Hawaii Island tended to begin and end on directionals and to change direction by shifting from one directional to another, and stream systems appeared to share much the same characteristics. On the Island of Kauai the hierarchical intervals showed up very clearly between stream courses. In chapter 4 additional examples from many parts of the world were presented to further confirm the correspondence of the directional network to the patterns of various fault and flow systems. The examples included coastlines from two large island groups and fault systems from the Indian Ocean, California, and Iran. The Laki flow from Iceland, the largest known historical flow, and the more recent pyroclastic flows from Mount St. Helens were also studied. The last three examples given showed the patterns of the three largest river systems in the world, the Mississippi, the Congo and Zambesi, and the Amazon. In all examples a close correspondence was found between the directional network and the pattern of the natural system. Based on the many examples, it seems possible that the strain analogy could represent a very simplified version of some physical process, so that the concept might prove useful in analysis even though incompletely understood. However, before discussing possible applications, an interesting diversion will be taken out beyond the Earth to other bodies in the Solar System that have been photographed over the last few years. Suprisingly perhaps, the directional pattern also seems to be reflected on the surfaces of these other worlds.

5 THE PLANETS

EXTENDING THE PATTERN

A geologic analysis is never routine and never really finished. Around the fringes of the analysis are the observational leftovers that refuse to fit into the particular selective process used. This tangled web of unused material always contains relevant bits and pieces of information that will probably find a place in the future analysis of a developing pattern. The process involves the never-ending, and often unconscious, sifting and sorting of the collected observations and ideas that form experience. But it is not experience itself that provides answers, rather the restructuring of the mental flux that searches for new, more inclusive patterns within the experience jumble. This search requires an active imagination, and perhaps even more important, a sense of wonder. Imagination is the mechanism used in the development of new mental patterns. It allows the mind to sort out and organize the diverse pieces of experience into coherent new forms. On the other hand, a sense of wonder provides the questing incentive to reach out and search the borders of experience for those oddities and anomalies that may form the basis for a new pattern. It is this wonder of the unknown world around us that drives people to the often boring work of collecting and classifying rocks, digging up fossils, tracing stream and fault networks, and mapping various Earth features.

Imagination supported by a sense of wonder has always extended beyond the Earth itself. High above was always the untouchable sky, filled with bright objects that had no counterpart on Earth and seemed a living, changing world of its own. The Sun moved across the daytime sky, bringing light and life to the world below while shifting back and forth across the horizon each year. The mysterious Moon shined dimly during the day and brightly at night. Over a month it gradually grew smaller as it moved toward the Sun until it finally disappeared, but then was born again to grow larger as it left the Sun until it filled out round and full at its farthest distance from the Sun. The Moon also moved back and forth across the horizon, but in monthly cycles in contrast to the Sun's yearly cycles. The Moon was undoubtedly woman, because the monthly phase cycle matched woman's menstrual cycle and nine Moon months matched woman's pregnancy period. The Moon measured time as reflected in the Indo-European root "me," from which came measure, metric, meal (time to eat), menstruate, month, and of course, Moon.

Overhead at night the stars were arranged in an unchanging pattern that moved across the sky and shifted through a complete cycle every year. Then there were the bright lights called planets that wandered through the pattern of stars. Mercury raced back and forth about three times a year between the Sun and the other planets, possibly carrying messages. Venus, much like the Moon, was the brightest of the wanderers when farthest from the Sun, but grew dim as it approached the Sun and finally disappeared when they merged. But Venus was reborn and grew brighter as it moved away from the Sun, until it became brightest just nine months later. Like the Moon, Venus must be a woman. Red Mars blundered and raged around his orbit, changing both his speed and brightness in what seemed an arbitrary way. Next came bright Jupiter who dependably completed his circuit in 12 years, neatly dividing the sky into 12 time slices, and last but not least was ponderous Saturn, farthest out and closest to the stars, who circled the heavens in 30 years in harmony with the Moon's 30 day cycle. Perhaps the clock in use today reflects Jupiter in its 12 hours and Saturn in its 60 minutes. Both 12 and 30 will divide evenly into 60. These are the five wondrous living planets that moved cyclically through the stars and spent their time measuring time.

Over the years science changed these living planets from Lords of time into inert physical entities circling the Sun along with the Earth, and more recently spidery space probes were shot from Earth to fly by the planets and their satellites and send back pictures. For some the wonder faded as these celestial bodies became tangible pieces of material, but for others the wonder merely changed to reflect somewhat different observations to be sifted, sorted, and analysed. Thanks to the foresight of the U.S. Government, and NASA in particular, the many photographs of the Solar System bodies have been compiled, integrated into maps, and made available to the general public. Surface maps of the Moon, Mercury, and Mars can be purchased just as easily from the U.S. Government Printing Office as maps of the Earth. Perhaps the biggest problem in using the planetary maps is the lack of supporting information to help interpret the physical meaning of the visual patterns. Tangible information is limited to a few rock samples brought back from the Moon and the simple soil analyses made by the two Mars landers. The limited information makes any analysis of the planetary bodies somewhat speculative, with the only real method available rather broad comparisons with the more understandable Earth. However, even a cursory view of the planetary maps reveal many linear features, and that raises the possibility that a directional pattern similar to that found on Earth might also be found on other Solar System bodies. The available information may be sparse and incomplete and the conclusions speculative, but the observations certainly include a sense of wonder.

5-1 Moon Mare Orientale

This shows the Mare Orientale Basin, located on the far side of the Moon so that it cannot be seen from Earth. As the Moon moves around the Earth its rotation acts to compensate, leaving the same side of the Moon always facing Earth. On the near side that can be seen from Earth there are several large darkened circular basins called maria or seas. This was the early name given the basins when telescopes were first used, and it was thought that the dark spots were large bodies of water. With the information collected by the Moon orbiters and landers, these basins are now known to be fields of lava that flowed out on the surface three to four billion years ago. Much of the lava pooled in huge circular craters, usually enclosed by concentric rings of mountain ranges that often rise several thousand feet above the surface. These multiringed basins are now interpreted as impact craters, formed when a large chunk of material slammed into the Moon and created a depression that later filled with dark lava. The pattern formed by the interconnected dark spots of lava resembles a face and is often called the "Man in the Moon."

The Orientale Basin on the far unseen side of the Moon is also a large multiringed impact crater, but it contains only a few small pools of lava that are shown as heavy lines on the figure. This lack of lava is characteristic of the far side of the Moon, which is very heavily cratered but has only a few patches of dark lava. It has also been determined that the crustal thickness on the far side is about double that on the near side, which may account for the heavier lava flows on the near side where the molten material was closer to the surface. Because the Mare Orientale crater was never filled with lava, details of the multiringed structure are relatively clear. The basin is defined by an outer ring of mountains with a diameter of about 600 miles called the Cordillera Range. Two small patches of lava lie within this range just east of the crater center. Inside the Cordillera Moun-

tain ring is another smaller concentric mountain range with a diameter of about 400 miles called the Outer Rook Range. Again on the east, an elongated patch of lava lies within the range. The center of the basin inside the mountain ranges is marked by a larger pool of lava.

A large multiringed crater such as Mare Orientale is formed by the impact of a very large body that releases an enormous amount of energy. It is estimated that the impact body that crashed into the Moon to form the basin may have been 100 miles across and traveling at 500 miles per hour or more. When impact occurs, a shock wave is created that is driven up into the impacting body and down into the Moon, and the pressure generated can be millions of times normal pressure. At impact the rock is compressed, blasted apart, and partially vaporized or melted, creating a rapidly expanding cavity with the rock material forced up and over the sides to spray out around the surrounding area. Some of the material blasted out of the crater piles up around the edge to form a lip, while other material spatters out and covers the immediate area with debris. The larger chunks form scattered small impact craters around the large central crater as they fall to the surface. At the point of impact the rock springs back as pressure is released, creating a central peak that sticks up through the pool of molten rock at the bottom of the crater. If the impacting object is very large, the central peak may be replaced by concentric rings of mountains that are probably shock waves that radiated out from the impact center and froze in place.

Surrounding the Orientale Basin is a radial system of lineations that splay out from the outer concentric Cordillera Mountain Range. Just as in a radial stream system, these rays do not appear to be spread out evenly like a fan, but rather tend to form clusters oriented along directionals. North of the basin a N7E trend is maintained across an interval of about 300 miles. Moving clockwise around the basin, the N7E trend stops rather abruptly giving way to a confused area on the east with no clearly defined trend. Con-

tinuing around the basin, a N40W trend becomes dominant to the southeast, again across an interval of about 300 miles. Further along to the south the trend shifts to N7E, with an area between these last two trend groups showing a mix of both. Moving around to the southwest, the N50E directional seems to be most evident, but with an overlay of N40W that gives an impression of small broken triangles. On the northwest the N40W trend seems to predominate before shifting back to the N7E trend north of the basin. At the basin center is a lava patch with the upper right and lower left sides trending N40W. These sides can be extended out beyond the Cordillera Range coincident with the N40W trend of the radial rays to the northwest and southeast. This same central lava patch is bordered on the right by a N7E directional that extends to the N7E radial ray groups on the north and south. A N40W trend may also be seen along the darker line running across the upper right corner of the page that represents the edge of a dark lava field on the near side of the Moon called Oceanus Procellarum. It seems somewhat surprising to see the Earth directional trends reflected on the Moon.

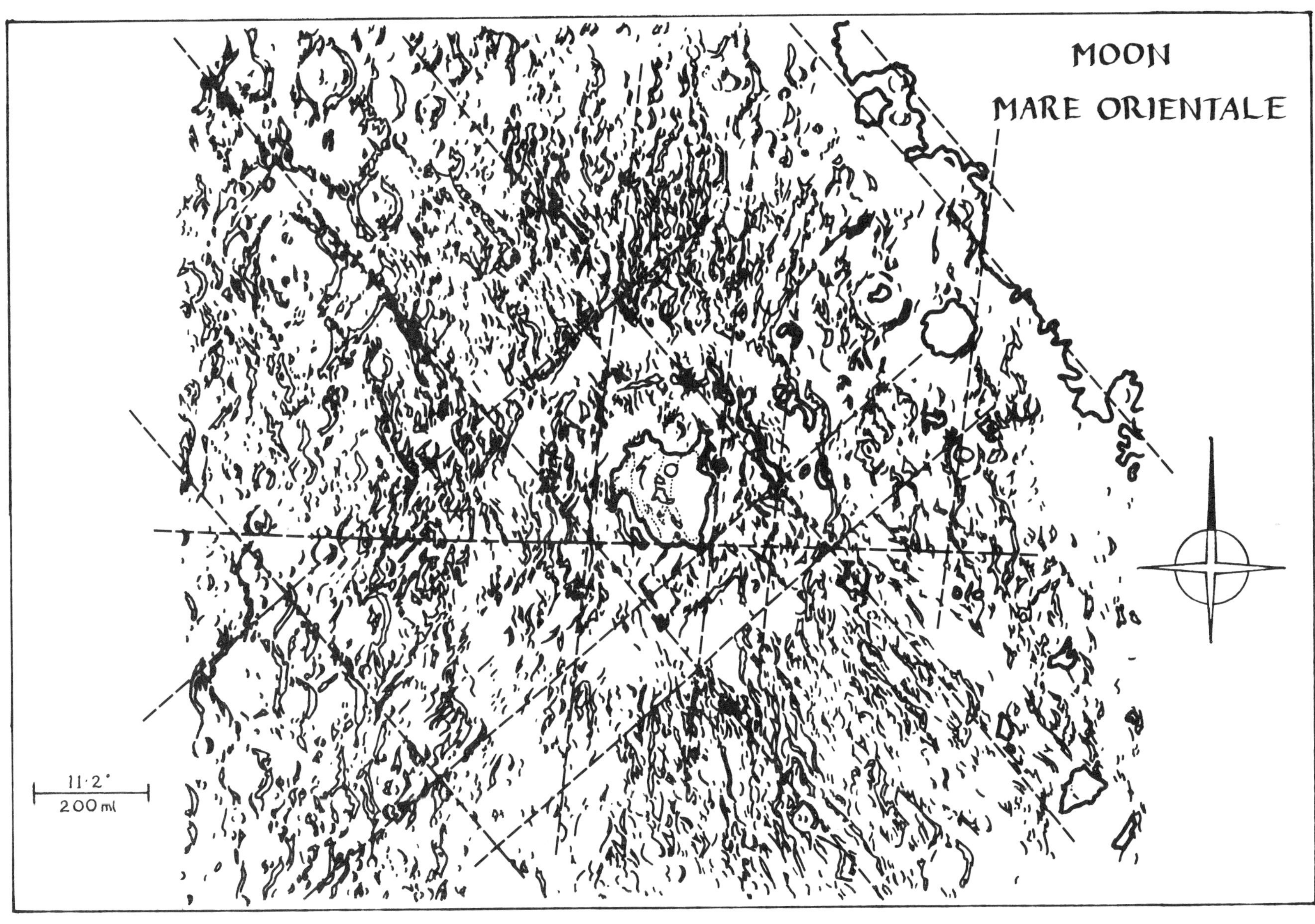
MOON
MARE ORIENTALE
11·2°
200 ml

5-2 Mercury Caloris Planitia I

This shows an area on the planet Mercury that includes a portion of the huge multiringed crater Caloris Planitia on the left. Only the eastern half of the basin was in daylight when the space probe photographed Mercury as it flew past, so that no information is available on the western half. Although the circular nature of the basin is not emphasized in this tracing, Caloris Planitia extends out about 400 miles from its center on the left edge, almost to the Odin Planitia on the east and Mozart Crater on the south, with its outer circumference marked by the Caloris Mountains. Farther east are a number of smooth plains separated by heavily cratered areas.

The word Mercury comes from the Etruscan "merc," relating to marketing or commerce, which was one of Mercury's special responsibilities in ancient times. He was also guardian of measurement and records, and invented writing, calculation, geometry, astronomy, and dice throwing, altogether quite an accomplishment. He was not only messenger of the Gods, but also messenger of death who lead the souls of the dead down to tartaros. However, he was certainly not perfect and had quite a reputation for craftiness, trickery, and stealing. All this was connected to a bright light in the sky that could only be seen just above the horizon at sunrise or sunset. Contrary to what some say, the planet can be easily observed above a relatively clear, unobstructed horizon when farthest from the Sun. The problem is where to look for it, and for those who know the star patterns this is no great problem. However, the planet is crafty, slipping quickly down below the horizon to hide most of the night, and dipping in and out of the Sun a bit more than three times each year. Mercury completes a cycle in 117 days, so that each 13 years it goes through 39 cycles. And the wonder is that 39 multiplied by 3 gives 117, so that the planet must be reckoning and computing as it sorts out the messages it gets from the Sun.

In Greek mythology Mercury is called Hermes and was closely associated with Apollo, who represented the Sun at dawn among other aspects. When Hermes was still a baby, he stole a herd of cows that the Sun had given Apollo. Hermes tied sandals to the cows' feet with the toes pointing backward, and then drove the cows into a cave to hide them. He killed two of the stolen cows and made a lyre by stringing seven cowgut strings across a tortoise shell. When Apollo finally found Hermes and the cows, he listened to the lyre and was so taken in by the music that he agreed to let Hermes keep the cows in exchange for the lyre. But crafty Hermes wanted more, so he tied seven reeds together to make a set of pipes and traded them to Apollo for a fiery gold staff which he used to herd the cows as he traveled to and from the Sun. The gold staff had originally been given to Apollo by the Sun. Meanwhile, the seven lyre strings and reed pipes played across a measured octave of time to tune the seven spheres of the universe as they shifted through their interwoven cyclic patterns. It is interesting that an eight whole note octave is made of seven tones with the first repeated at the scale end, and if both whole and half notes are used, the scale has thirteen notes made of twelve tones, again with the first repeated at the scale end. It is as if a circle had been marked with seven or twelve equal divisions, and then cut apart at a division to add one more mark at the snipped end. Earthly music measures out the heavenly time cycles of the seven wandering lights in the sky, the seven days of the week, the twelve months in a year, and the twelve years in a Jupiter cycle. Of course, the heavenly interactive clockwork analysed by our ancestors was far more complicated than these simple relationships. People have been analyzing the world for many thousands of years, and perhaps the mental processes have not changed as much as we choose to believe.

Before the space age, very little was known of Mercury, because telescope observation was very difficult with the planet so close to the sun. Then in the mid 1970's Mariner 10 flew past the planet three times, sending back pictures and measurements. This provided information on Mercury similar to that available on the Moon before the Apollo program of space probes and Moon landings. The surface photographs cover about half of Mercury with a resolution comparable to that of the Moon through an Earth based telescope. The pictures show a heavily cratered surface resembling the Moon, with a number of intercrater plains that are scattered across the surface rather than grouped together as on the near side of the Moon. The Mercury plains are probably volcanic, although the evidence is inclusive. The cratered areas contain many linear features that can be seen on the eastern portion of the figure. These features are predominantly oriented N50E, but also with a N40W overlay grain. The result is a pattern that gives an impression of a jumble of relatively small broken rectangles with sides trending along the diagonal directionals. The pattern is also reflected in the boundaries of the plains which appear almost square in shape. In the next figure an enlarged portion of the lower right quarter of Caloris Planitia and the surrounding area will be presented.

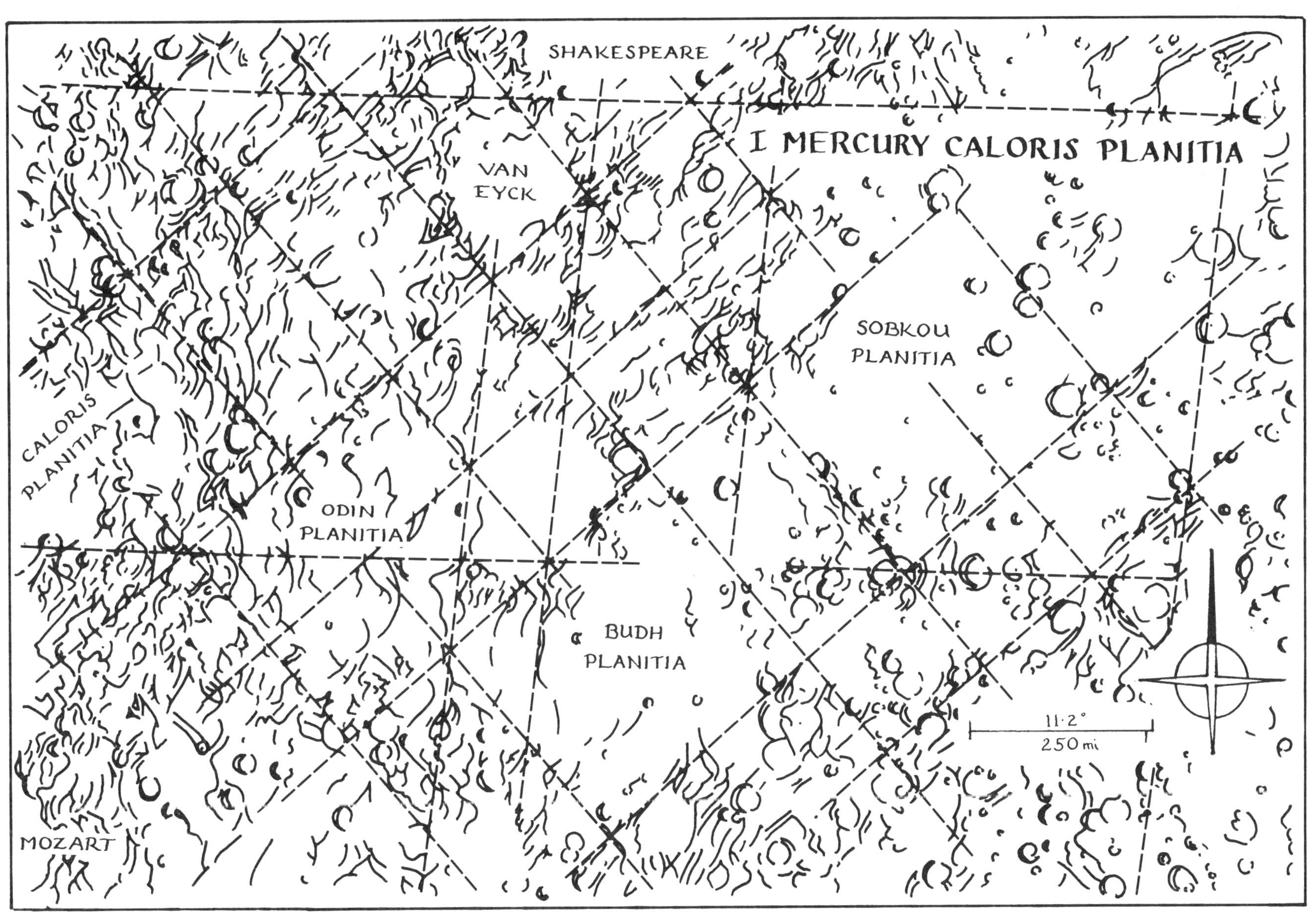
SHAKESPEARE
I MERCURY CALORIS PLANITIA
VAN EYCK
SOBKOU PLANITIA
CALORIS PLANITIA
ODIN PLANITIA
BUDH PLANITIA
MOZART
11·2°
250 mi

5-3 *Mercury Caloris Planitia II*

This shows a portion of Caloris Planitia in the upper left corner of the figure. The circular mountain range, called the Caloris Mountains, that defines the outer boundary of the basin lies on a quarter circle between 300 and 400 miles out from the upper left corner. The central portion of the basin in the extreme northwest is crisscrossed with many fractures and ridges forming a pattern of generally rectangular polygons. The width of the fractures in the basin interior varies from about one-half mile, the lower limit of resolution, to a maximum of about five miles. The wider fractures appear to be about one-half mile deep, and the ridges between 1,500 and 2,000 feet high. The fractures have a flat floor, indicating that the surface has probably dropped down between two parallel faults, forming a valley called a "graben." No features of this type have been found in the multiringed basins on the Moon or Mars, and there is no adequate explanation why the grabens are found in the Mercury basins.

The circular Caloris Mountain Range surrounding the basin appears on the figure as a somewhat jumbled pattern of lines and craters. Out beyond the mountains the pattern resolves into rectangular sets of lines representing long hills or ridges that are especially prominent in the vicinity of Odin Planitia. The entire area outside the Caloris Basin is probably covered with debris sprayed out from the crater when the impact occurred. The lines forming the rectangles both inside and outside the basin generally trend along the N40W and N50E Earth directionals, with a faint indication of the hierarchical intervals within the pattern. the overall impression is somewhat different from the examples found on Earth, where individual features often do not follow directionals, but rather begin, end, or change direction on a directional. This difference between the patterns on Earth and Mercury could be merely a visual effect. Small individual features less than about one-half mile in size cannot be seen on the Mercury photographs because of insufficient resolution. The features that can be seen are much more complicated than they seem, representing only a rough visual impression of a multitude of smaller unseen features grouped together in some sort of pattern. An analogy might be the visual impression of canals on Mars when the planet is viewed through an Earth based telescope, and the smaller features can only be seen as a shadowy network. Space probe photographs of Mars with much higher resolution show a considerably more complex surface pattern, but no canals. In the Earth examples the N40W and N50E directionals are most clearly defined, representing peak deformation along the diagonals of the strain rhombus. It would seem that these diagonal directionals would also show up most clearly in a gross visual impression of many smaller features.

The strain analogy is consistent with the patterns of directionals found around the multiringed basins of Mare Orientale on the Moon and Caloris Planitia on Mercury. As described in the last chapter, a multiringed basin is formed by the impact of a relatively large object striking a larger body. The impacting object carries an enormous amount of energy because of its large mass and high velocity. When the object crashes, it is stopped short almost instantaneously, and the energy it carries is changed to a huge expanding shock wave that travels through the body that was struck, and on to the opposite side where additional damage is centered. The damaged surface on the other side of the planet opposite the impact area of Caloris Planitia shows up as a limited area of rough jumbled topography and many broken crater rims. On the Moon opposite Mare Orientale there is a similar area that appears to have been torn up by an impact shock wave. Both the physical impact and the impact shock wave tend to rip apart the surface of the body that is struck. If the surface of that body is in torsional strain when the impact occurs, it might be expected that the resulting fracture pattern would reflect the strain pattern in some way. This could explain the existence of the directional patterns found around the multiringed impact basins on the Moon and Mercury.

The extreme temperature differences on Mercury may also play some part in emphasizing the directional pattern. Mercury is not only close to the Sun, but has almost no atmosphere to shield the planet, so that the full radiation of the Sun falls on the exposed surface. Mercury spins around its axis as it circles the Sun, and the speed of these motions are syncronized, creating two hot spots on opposite sides of the planet that receive about twice the average sunlight intensity. Perhaps coincidentally, these hot spots are centered on Caloris Planitia and on the opposite side where the impact shock wave caused the damaged area. Temperatures in these hot spots vary between 800 degrees Fahrenheit during the day and -300 degrees Fahrenheit at night, a difference of 1,100 degrees. This extreme temperature difference causes a cyclic expansion and contraction that acts to break up the surface, and the resulting fracture pattern would tend to reflect in some way a strain pattern within the surface.

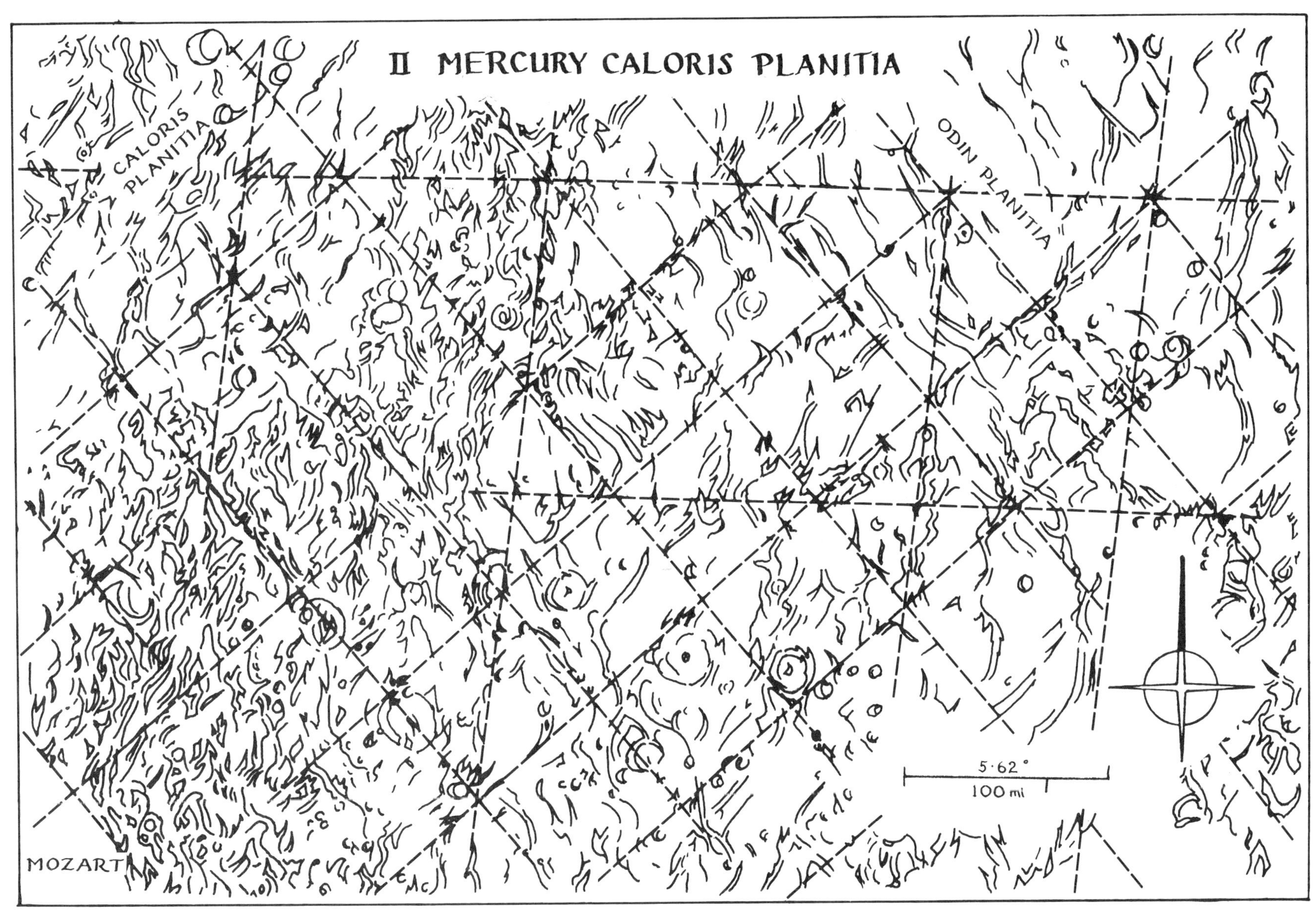
II MERCURY CALORIS PLANITIA
CALORIS PLANITIA
ODIN PLANITIA
MOZART
5·62°
100 mi

5-4 Mars Noctis Labyrinthus

This shows the western end of an enormous canyon system, called the Valles Marineris, that stretches along the Martian equator for 2,500 miles. The system is 400 miles wide in some places, with single canyons up to 125 miles wide and 40 miles deep. In comparison, the largest canyon on Earth is the Grand Canyon of Arizona which is only 300 miles long and 20 miles across. The west end of Valles Marineris, called Noctis Labyrinthus, is a geometric network made up of many intersecting canyons that are relatively short and narrow. These narrow canyons appear to be unconnected rows of flat bottomed pits of varying depth. They may be grabens similar to those found in Caloris Planitia on Mercury that were formed by the collapse of the surface between pairs of parallel fractures. Noctis Labyrinthus is on the crest of a huge blister called the Syria Rise, or Tharsis Region, where the surface of Mars has bulged up. This roughly circular blister which straddles the equator is about 3,000 miles across, a distance equivalent to one-third of the planet's circumference. Several of the largest volcanoes of Mars are located on the blister which rises up about four miles above the Mars' surface. It may be that Noctis Labyrinthus on the Tharsis bulge is a fracture system formed as the growing blister stretched and tore the surface.

In the lower right quarter of the figure the network of canyons is roughly bounded by a "V" with sides that follow a N40W directional on the west and a N50E on the east. Just southeast of the "V" apex the canyons form a diamond with sides also oriented N40W and N50E, and within the "V" the intersecting canyon network also reflects the same diagonal directionals. However, just above the "V" apex a canyon follows a N7E directional that can be extended to the extreme north where it forms the eastern boundary of a large group of parallel N7E fractures running across the top of the figure. Moving up the west side of the "V," the N40W directional shifts abruptly near the center of the figure to W2N.

Further west in the left half of the figure, the canyons merge to form a complex fracture pattern that strikes off between two N50E directionals, and then gradually curves down to end up along N40W in the extreme lower left corner. Throughout the overall pattern there is a reflection of the hierarchical intervals, although the pattern complexity confuses the picture. Again, as with the Moon and Mercury, it is somewhat surprising to find the Earth directionals appearing on the Mars' surface.

Mars has always been a prime subject for speculation. The ancients always considered the planet a central sky figure, located as it was between the inner planets of Mercury and Venus that hugged the Sun and the outer planets of Jupiter and Saturn that slowly circled the sky measuring the cycles of time. Mars swings around the Sun just outside the Earth's orbit, so that the brightness of Mars varies considerably. When Mars is on the opposite side of the Sun from the Earth, it is far away and faint, but when it closes with Earth on the same side of the Sun its diameter increases up to seven times, and it becomes very bright. This cycle of brightness takes a little more than two years, while its cycle through the stars is a little less than two years. But the brightness cycle is further complicated because the distance between Earth and Mars can vary considerably when they are close together on the same side of the Sun at opposition. The total effect of all these differences gives a complicated schedule in which Mars shines brightest through a 79 year sequence at intervals of 15,17,15,17, and 15 years, which then repeats. It is no wonder that Mars was known as impetuous, illogical, and destructive.

Mars was called "Ares" in Greek and was the son of Zeus and Hera. Both Ares and his twin sister Eris loved war with its killing and destruction, with no real concern of who fought or who won. He was known for his blind and unreasonable anger and was often represented as a wild boar, the form he took when he killed Adonis in a fit of jealousy. He was a bull in a china shop who blundered through the stars along the flaming lava river of Pyriphegethon, following his complicated path of cycle on cycle. But for all his destruction, there was always a certain admiration, or perhaps pity, for Ares, as we all have for someone ruled by impulse and feeling rather than logic and rationality. In a sense, he represents the brave and unthinking warrior who rushes out to fight regardless of the circumstances or odds. If he is lucky and wins, he is the acclaimed hero, but if his luck runs out or he picked the wrong side, he becomes the notorious brute, killer of innocents, and eternally damned. For the Greeks he represented that unfair, destructive, and unpredictable aspect of nature that rational science would just as soon forget. Nature's destruction by earthquake, volcano, landslide, hurricane, or tsunami is not selective, and there is an end for all, young and old, good and evil, innocent and guilty, wise and ignorant, rich and poor, and both mighty and low. But it is the way of the world, and one way or another there is often little recourse but acceptance.

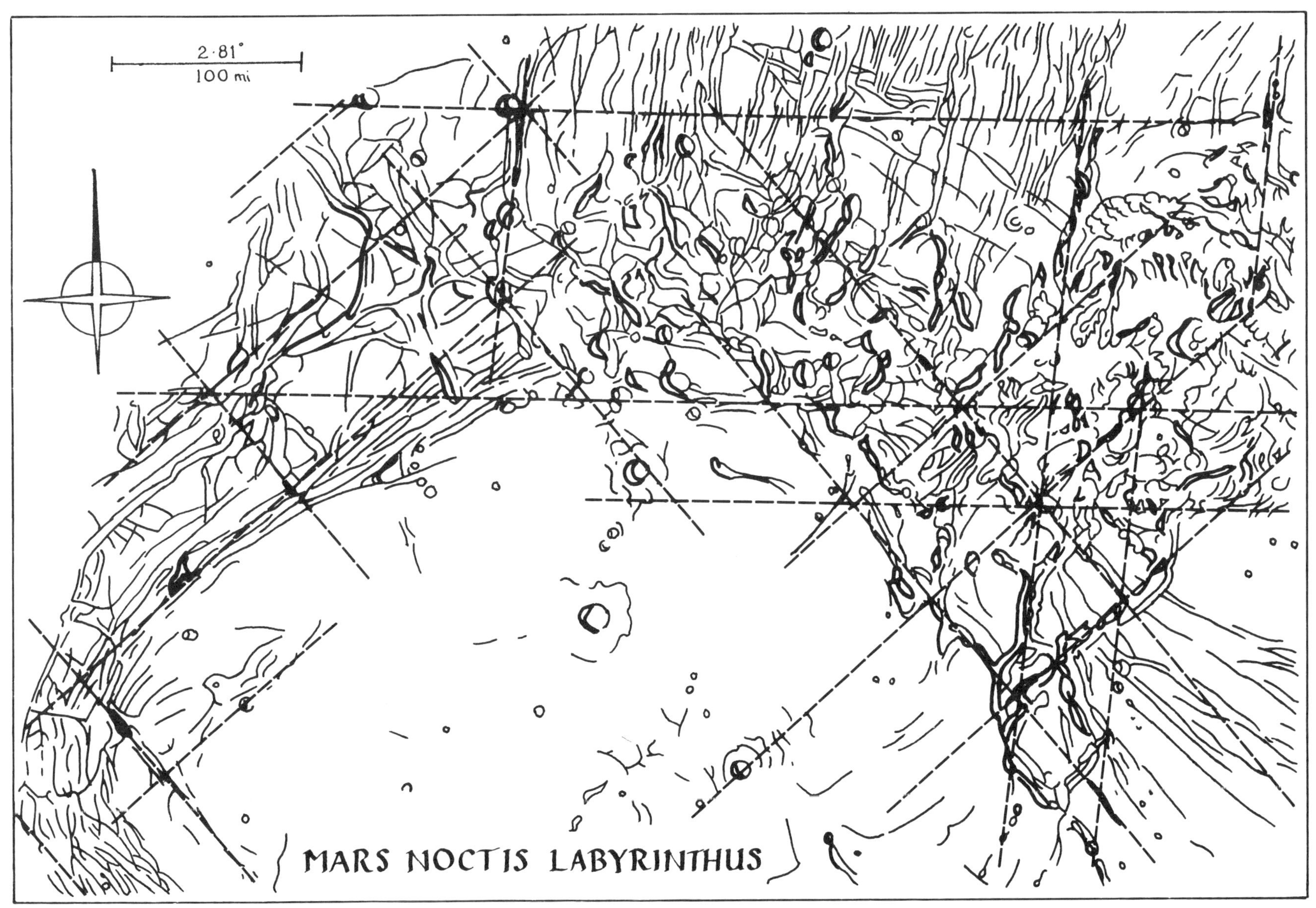
2·81°
100 mi
MARS NOCTIS LABYRINTHUS

5-5 *Jupiter Satellite Callisto*

This shows the remains of a large impact crater on Callisto, one of Jupiter's four Galilean satellites. These four large moons were first seen by Galileo in January 1610 when he first began his observations through the newly invented telescope. They are named Io, Europa, Ganymede, and Callisto, all lovers of Jupiter, or Zeus as he was known to the Greeks. Both names can be traced back to the Indo-European root "dei" which refers to the bright, clear sky where the gods lived. The word Jupiter comes from "Zeus-peter," or "Zeus the Father," a reasonable designation for the brightest planet in the sky, except for Venus. But Venus only appears low in the sky in the evening or morning, and often fades away completely, while Jupiter shines steadily through much of the night.

The planet Jupiter is enormous with a mass greater than all other planets and satellites combined. With its fifteen known satellites, it forms its own miniature solar system, and is considered by some astronomers to be part of a double star system made up of Jupiter and the Sun. It moves purposefully through the stars, clicking off twelve years each time it circles the sky, and comes back to the same spot in the star pattern after completing seven circles over 83 years. These two numbers, twelve and seven, are central to many of man's concepts. The measure of time is based on twelve, with a day twice twelve hours and a year twelve months, corresponding to the Zodiac of twelve signs. Produce items are often sold by the dozen, and the musical chromatic scale contains twelve half tones. Court cases are decided by a jury of twelve. Seven is perhaps even more extensive. There are seven days in a week and seven colors in a rainbow. Oceans are divided into seven seas and there are seven tones in a scale's eight note octave. There are seven wandering objects in the sky, including the Sun and Moon, seven celestial spheres, and seven bright stars in the widely known constellation often called the "bear." Both heaven and hell have seven levels and there are seven deadly sins. It is interesting how the concepts of the past carry on through the centuries.

Before the Pioneer and Voyager space probes, many physical characteristics of Jupiter were known from telescope observations. Size and shape could be directly measured, and mass was computed from the circular paths traced out by the satellites. Atmospheric composition was estimated from spectrographic analysis, and infrared observations gave an estimate of temperature. Jupiter was found to be made up mostly of hydrogen and helium gas, much like the Sun. The visible surface was banded with a complex of shifting, multi-colored belts that circled the planet. The many different colors of the bands indicated that they must contain minor amounts of many elements in many complex forms. Imbedded in the colored bands was the Great Red Spot that had stayed in about the same place ever since it was discovered in 1664 by Robert Hooke. The temperature of the cloud tops on Jupiter was very cold, measuring about -350 degrees Fahrenheit, but gradually warming at greater depth. The shape of the planet was oval, slightly squashed at the poles because of its rapid spin. Four bright lights were observed close-by that continually shifted position and were correctly interpreted by Galileo to be four large satellites circling the planet at different orbits and speeds.

These four Galilean satellites were tracked by many observers over many years, so that eventually the pattern could be predicted quite accurately over time. Jupiter and its satellites could then be used as a giant clock, keeping exact universal time overhead in the sky. This gave a solution to the major surveying problem of determining longitude, which required a clock that gave the same time anywhere in the world. Captain Cook, one of the world's greatest navigators and surveyors, spent much of his first few months in the Pacific on Tahiti, observing the changing pattern of Jupiter's satellites in order to precisely set his ship's clocks. The satellite pattern was also used in the first accurate determination of the speed of light.

As Jupiter recedes from Earth, the satellite clock runs slower, but then speeds up as the planet approaches Earth. The Danish astronomer Ole Roemer concluded that the difference was due to the changing distance that light must travel from Jupiter to Earth, and used the concept to compute the speed of light.

With the Voyager fly-by in 1979, these four points of light finally became real worlds, each with its own unique characteristics. Nearest Jupiter is bright, reddish Io with a surface molded by volcanic activity, and just beyond Io is Europa, a smooth icy snowball crisscrossed with a network of thin dark lines. Ganymede is next, with mysterious intersecting grooved bands interspaced with heavily cratered areas. Farthest out is Callisto, the most intensely cratered body yet found in the Solar System. The most prominent feature on Callisto is Vahalla, the fossil image of what was probably a huge multiringed impact crater. It has a very smooth surface, with the pattern defined by the bright center and surrounding circular ripples. The bright center is outlined by a square with 22.5 deg sides orientated N40W and N50E. The west corner of the central square is cut off by a N7E directional that can be extended both north and south. All four directionals tend to trace changes and breaks both in the surrounding ridges and further out in the faultlike network of broken linears.

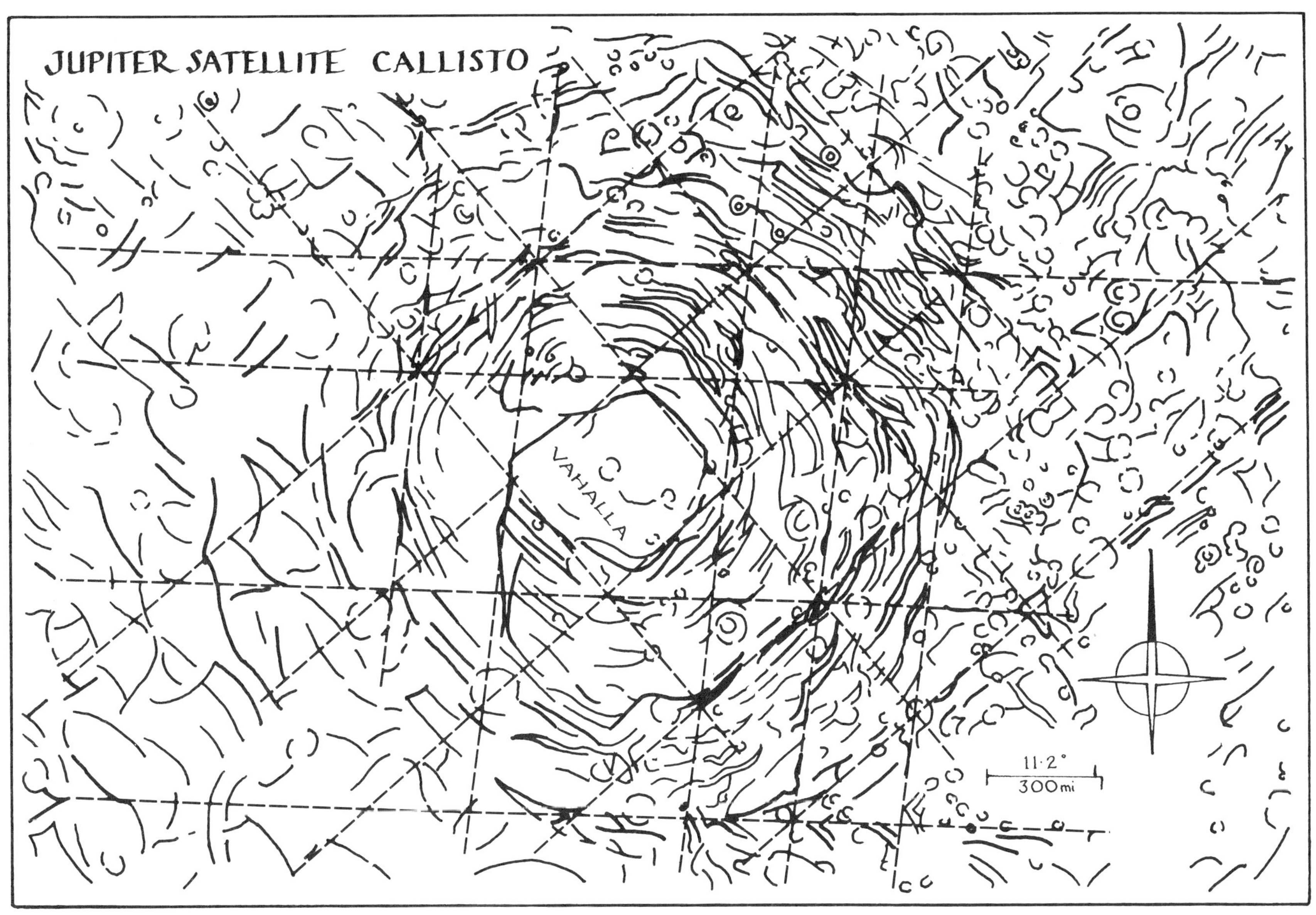

JUPITER SATELLITE CALLISTO
VAHALLA
11·2°
300mi

5-6 *Saturn Satellite Enceladus*

This shows Enceladus, one of the 17 satellites of Saturn. The satellite was photographed during the Voyager fly-bys of 1980 and 1981. The planet Saturn has always held an important place in cosmic concepts. It is farthest out and faintest of the five visible planets, and moves most slowly along its path through the stars. Saturn circles the sky in 29.5 years, a reflection of the 29.5 days it takes the Moon to go through a phase change. Both Saturn and the Moon were thought of as the essence of time. As mentioned before, the word Moon comes from the Indo-European root "me," meaning measure or setting of time, and most early calendars were based on the month as marked off by the Moon. The Greeks knew Saturn as "Khronos," sometimes spelled Cronos, which is just another word for time. The word Khronos comes from the Indo-European root "gher," which means to seize, embrace, or enclose, as time itself embraces all the universe as an integral whole. Saturn was Lord of all measures and originator of time as he steadily moved through the stars and handed out measures of time to the other planets as they sped by him.

Saturn has seven major satellites that include Titan, the largest, and Enceladus, one of the lesser six. Enceladus was first seen by William Herschel, probably the greatest astronomer of the 18th century. It is largely made of water ice, and is the brightest and whitest of all the Saturn satellites. Its surface is covered with scattered areas of heavily cratered patches, corrugated linear features, and smooth plains. The complex surface is an indication that deformation of some kind has probably effected the surface fairly recently. This is somewhat surprising, considering the extremely cold temperature of the icy surface, estimated at -350 degrees Fahrenheit, which would make it very rigid. Saturn is a long way from the Sun and receives much less radiation than the closer worlds. It may be that changing gravitational forces acting between Enceladus, Saturn, and nearby satellites would continually squeeze the icy interior. The heat generated could partially melt the ice and gradually force the liquid upward to break out at the surface as an ice volcano. However, there is a great deal of speculation in any explanation, because there is little to go on except surface photographs. But what a wonder that cameras could be carried far out in space to send back pictures of objects that are only points of light as seen from Earth.

The surface pattern of Enceladus shows a network of very clear linear trends. As with many examples, the N40W and N50E directionals stand out most clearly, consistent with the strain analogy in which the rhombus diagonals represent lines of peak distortion. Several directionals are well defined in the vicinity of the clear white patch in the center. The white patch is cut into two parts by a N50E dividing directional. The lower southeast part of the patch is rectangular, with the upper left side of the rectangle coincident with the dividing N50E directional, and the lower right side also on a N50E directional lying 22.5 deg to the east. Both upper and lower parts of the white patch are bordered on the upper right by the same N40W directional. Just across this directional, northeast of the lower part of the patch, is a rectangular area containing several well defined diagonal directionals. The dividing N50E directional and the N40W directional forming the upper right border of the patch can both be extended upward to form the sides of a right triangle with its apex pointing down. The base of this triangle lies along a W2N directional that can be traced eastward almost to the right edge of the figure where the linears turn down abruptly along a N7E directional. The linears in the W2N triangle base can also be extended a short distance west where they again abruptly turn down along another N7E directional that borders the west side of the central white patch. These three directionals form the upper part of a frame with its top running W2N along the triangle base, and sides following the two N7E directionals. The linears west of the white patch form a band lying between two N7E directionals just 22.5 deg apart. Southwest of this N7E band the grain shifts to N50E, forming another band of linears lying between two N50E directionals 45 deg apart. The lower N50E directional bordering the band, when extended, is the same directional that divides the patch and also forms the lower right side of the right triangle described above. Another interesting observation is the indication of nodes within the pattern of linears where the directionals intersect.

The occurrence of Earth directionals on satellites of other planets in the Solar System would indicate that the force causing the effect was in some way related to the Sun, which dominates all objects in the system. It is interesting that the north trending directional when computed from hierarchical concepts is north 7.2 deg east, which just happens to be the declination of the Sun to the ecliptic. Although this may be purely coincidental, analysis looks on coincidence as highly suspicious.

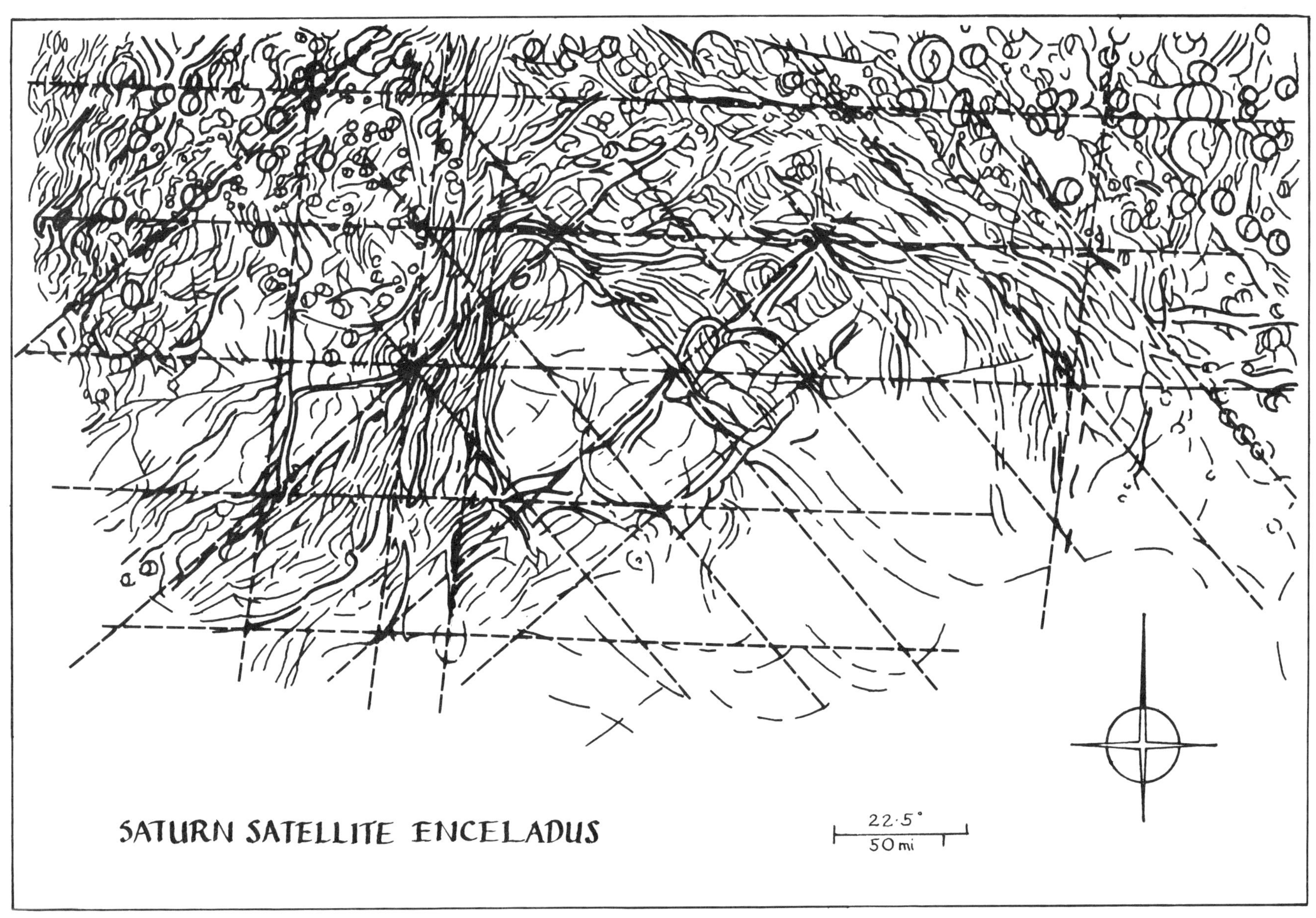
SATURN SATELLITE ENCELADUS
22.5°
50 mi

5-7 Saturn Satellite Tethys

This shows the surface of Tethys, another satellite of Saturn. In ancient Greece, Tethys was one of the seven Titans who were children of Mother Earth and Uranus the sky. Saturn, Cronos in Greece, was the youngest Titan of the seven. As with many families, the spoiled Titan children grew to resent their father ordering them about, and became nasty about the whole thing. However, their reaction was much more drastic than that of most children, and with Mother Earth urging them on, Cronos led them into battle against their father Uranus. Armed with a flint sickle his mother had given him, Cronos castrated his father and then threw him down to the bottom of Tartaros, the timeless land.

Cronos made the first cut in the circle of time, but it took a second cut to separate an interval of time and start the clock ticking. After getting rid of his father, Cronos took charge and ruled the timeless "Golden Age" when there was no evil and where he walked and talked with people. In those days, the people lived on milk and honey and there were no cares or troubles. Of course, without time, death had no meaning and was no more than sleep. But history has a tendency to repeat itself, especially within families, and Cronos' children turned out much the same as the children before. Cronos thought he had solved the problem by swallowing each child when it was born, but his wife didn't think much of the idea. So when Jupiter was born, known as Zeus to the Greeks, she fed Cronos a stone instead of the child. When Zeus had grown to a young man, he came back and tricked his father Cronos into vomiting out the other children, who were not only alive but, wonder of wonders, had become full grown inside their father. The children turned against Father Cronos and finally defeated him in a conflict that shook the Earth and split off the first interval from the broken circle of time. Father Cronos, like Father Uranus before him, was also thrown down into timeless Tartaros. But the essence of Cronos continues to circle the sky above as

Saturn, marking out 12 deg each year and completing a full circle through the stars every 30 years.

Saturn is best known for its beautiful ring system that makes it unique among the planets. Other planets have rings, but none can compare with the intricate and complex Saturn ring system. It was only recently in 1980 and 1981 that the full complexity of the system became known from pictures taken by the Voyager spacecraft. The system is a thin disk like a hat brim made of small particles that circle the planet from about 4,500 miles above the surface out to 45,000 miles. Within the disk are thousands of narrow ringlets, probably representing waves traveling through the particle clouds. Each ringlet has a distinctive width, brightness, and color, with ringlet groups isolated by a few wide gaps almost empty of particles. The particles that make up the rings are mostly pieces of water ice that range in size from dust to large boulders. All this combines to make Saturn and its rings one of the most fascinating objects in the heavens.

Tethys is the third satellite out from Saturn, and circles the planet just beyond Enceladus. Its diameter is a little more than 600 miles, about twice that of Enceladus, and like Enceladus, is essentially a giant snowball made of water ice. Tethys was first seen by the great 17th century astronomer Jean Dominique Cassini, who was also first to observe the large gap in Saturn's rings which is still called the Cassini Division. Like many outer planet satellites, much of Tethys is heavily cratered, indicating a very ancient surface. However, like other satellites photographed by Voyager, Tethys has its own unique characteristics. One side of the satellite is dominated by a huge crater about 250 miles across which can be seen in the upper left of the figure. The crater surface is relatively smooth with little relief, and is probably an old impact scar that has been healed by plastic flow on the icy surface. On the opposite side of Tethys there is an enormous complex of valleys and trenches over 1,000 miles long that can be seen running from top to bottom near the right edge of the figure. It

seems probable that this great crack was made by the shock wave from the impact on the opposite side that caused the crater scar. It will be remembered from discussions of multiringed craters on the Moon and Mercury, that when a large object hits a body, a tremendous shock wave is generated that travels through the body and disrupts the opposite side. A disruption of this magnitude could easily have caused the extensive fracturing that forms the valley complex.

A number of directionals can be traced on the satellite surface. The bottom of the crater edge is a "V" with the left side running N40W and the right side N50E. Just 22.5 deg northeast of the left side is another N40W that crosses the crater. The east edge of the crater is bounded by a N7E directional. As with many fault systems, the valley complex on the far right does not trend along a directional. However, very distinct linear trends can be seen in the disturbed area lying between the valley complex and the crater. The principle trend within the area is N50E, with the directionals reflecting the hierarchical intervals. In addition, there are faint traces of N40W directionals also crossing the area.

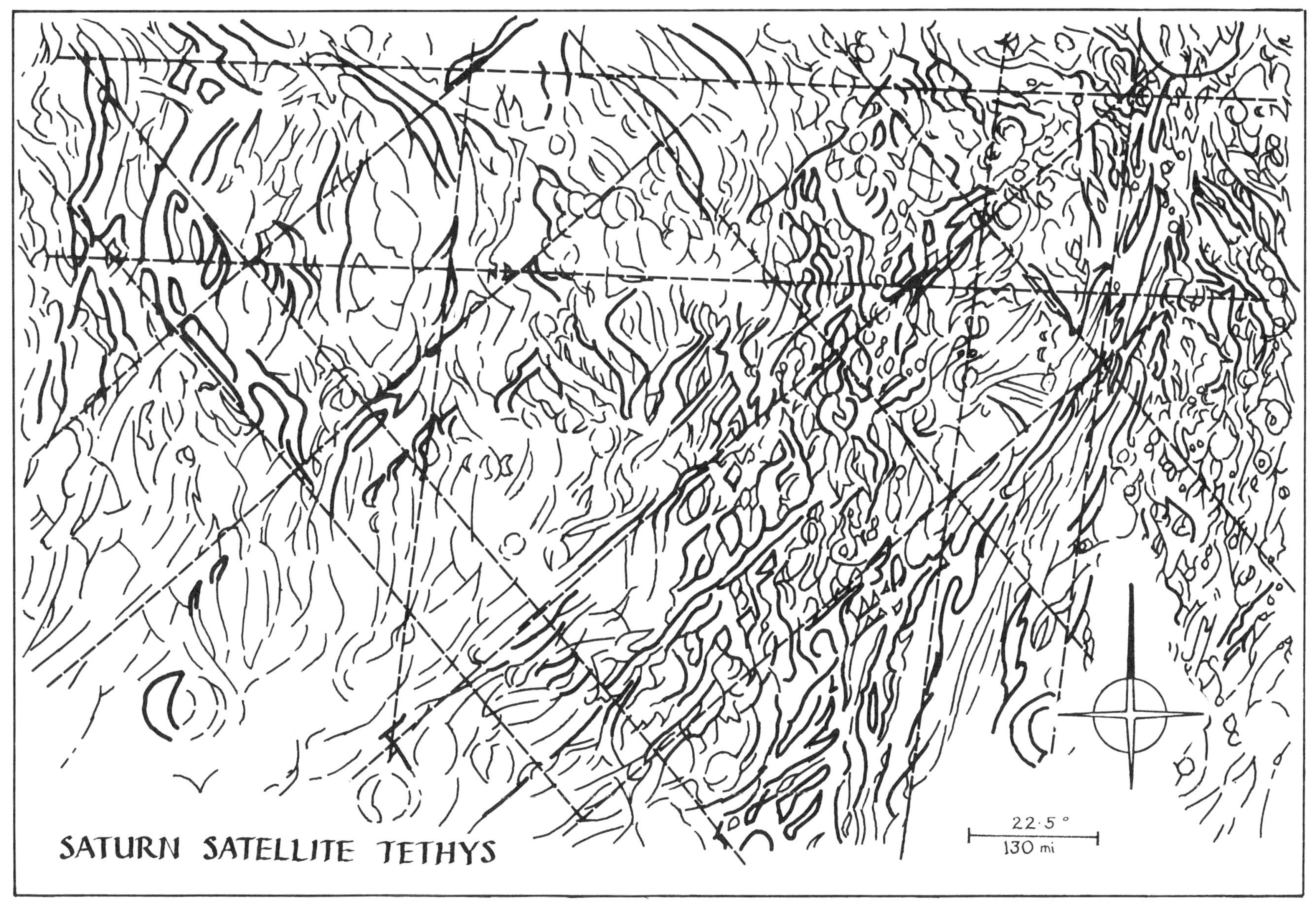

SATURN SATELLITE TETHYS
22.5°
130 mi

5-8 Jupiter Satellite Europa

This shows the surface of Europa, one of the Galilean satellites that circle Jupiter. In ancient Greece Europa was the granddaughter of Poseidon, known to the Romans as Neptune. Poseidon controls the sea and is associated with earthquakes and horses, and when he rages the storm winds scream and huge waves rise up to batter the ships and flood the coast. Of course his granddaughter was beautiful, and when Zeus saw her he immediately fell in love. This was not unusual for Zeus who had a tendency to run after every pretty girl he saw, using every trick in the trade to seduce the lady. He often used a disguise to fool the girl he wanted, and this time changed himself into a snow-white bull with a black streak between his small polished horns. Europa was so taken with the beautiful creature that she made the mistake of climbing up onto his back. Zeus in the form of the snow-white bull quickly ran into the sea and swam away with Europa to Crete where she bore him two sons. Perhaps the moral of this story is that no pretty girl should climb up on the back of a snow-white bull, or perhaps not. Europa the satellite has turned out to be as wonderful as the story of Europa the Greek maiden. The surface of the satellite is made entirely of ice and is as smooth as a billiard ball. It is covered with an intricate network of fine intersecting lines, giving the appearance of a cracked egg. Most of the lines are grey streaks that crisscross the smooth surface with no indication of any indentation. A few lines are white ridges about 1,000 feet high that form a scalloped pattern which repeats every 50 or 100 miles. Very few impact craters have been found, indicating a fairly young surface that has been disturbed by some sort of internal activity. No satisfactory suggestion has yet been made to explain the network pattern of fine lines etched on the smooth surface.

Pictures of Europa were taken on the Voyager mission that was conceived in the late 1960's before man had stepped out on the Moon. At intervals of about 180 years the four outer planets assume a unique configuration that allows a single spaceship to visit them with a relatively small amount of fuel. This is done by using a planet's gravity to change the course of the spacecraft as it passes, and throw it on to the next planet where the process is repeated. NASA began to design a craft that could start its trip from Earth in the late 1970's when the outer planets would move into that unique configuration. Based on funds, time, and technical constraints, two identical spacecraft called Voyager one and two were designed to visit Jupiter and Saturn, with the possibility that one might go on to Uranus and Neptune if all went well. As it turned out, the mission went even better than well, and all objectives have been met. The Voyager craft looks like a large spider with a round saucerlike body and five long legs. It is about as big as a car and weighs about a ton. The saucerike body is a dish antenna that transmits the observations back to Earth, and the spider legs hold the instruments that collect the information. Both Voyagers have visited Jupiter and Saturn, and Voyager two flew by Uranus in 1986 and Neptune in 1989. This completed the "Outer Planet Grand Tour" that has turned out to be one of the greatest technical achievements ever accomplished.

Although the network of lines on Europa has yet to be explained, it shows many of the characteristics of fracture systems on Earth. A few of the major lineaments trend along directionals, but most head off in other directions, with directionals marking the lineament ends and direction changes. As in an Earth fracture system, the directional pattern provides a network that frames and supports the more complex natural system of fractures. The hierarchical directional system is the organizing structure on which the individual elements of the natural system hang. In the left half of the figure the prominent Asterius Linea follows a N50E directional for more than 1,500 miles. This forms the upper boundary of a band that is crisscrossed by a number of shorter lineaments. Within the band are three 22.5 deg wide strips that are defined by three more N50E direc-

tionals which are coincident with several short but prominent lineaments in the upper center of the figure. At about the center of the page, the N50E band butts up against a central N7E directional, where the upper lineaments begin to curve over towards the east and down. However, the lower boundary of the band shifts direction abruptly from N50E on the left to N40W on the right of the central N7E directional.

There are several prominent sets of lineaments east of the central N7E directional. Towards the top of the page the lineaments tend to curve around and down, until they straighten out along N40W directionals. At center right there is another, smaller group of prominent lineaments that form a complex of polygons. The group is bounded on the northeast and southwest by two N40W directionals separated by a 22.5 deg hierarchical interval. These directionals form the two longer sides of a rectangle. The short upper left side of the rectangle follows the lower N50E boundary of the band lying below Asterius Linea, and the lower right side follows a very prominent linear in the lower part of the figure that trends along another N50E directional. Throughout the figure, directionals can be observed that follow breaks in the less prominent trends.

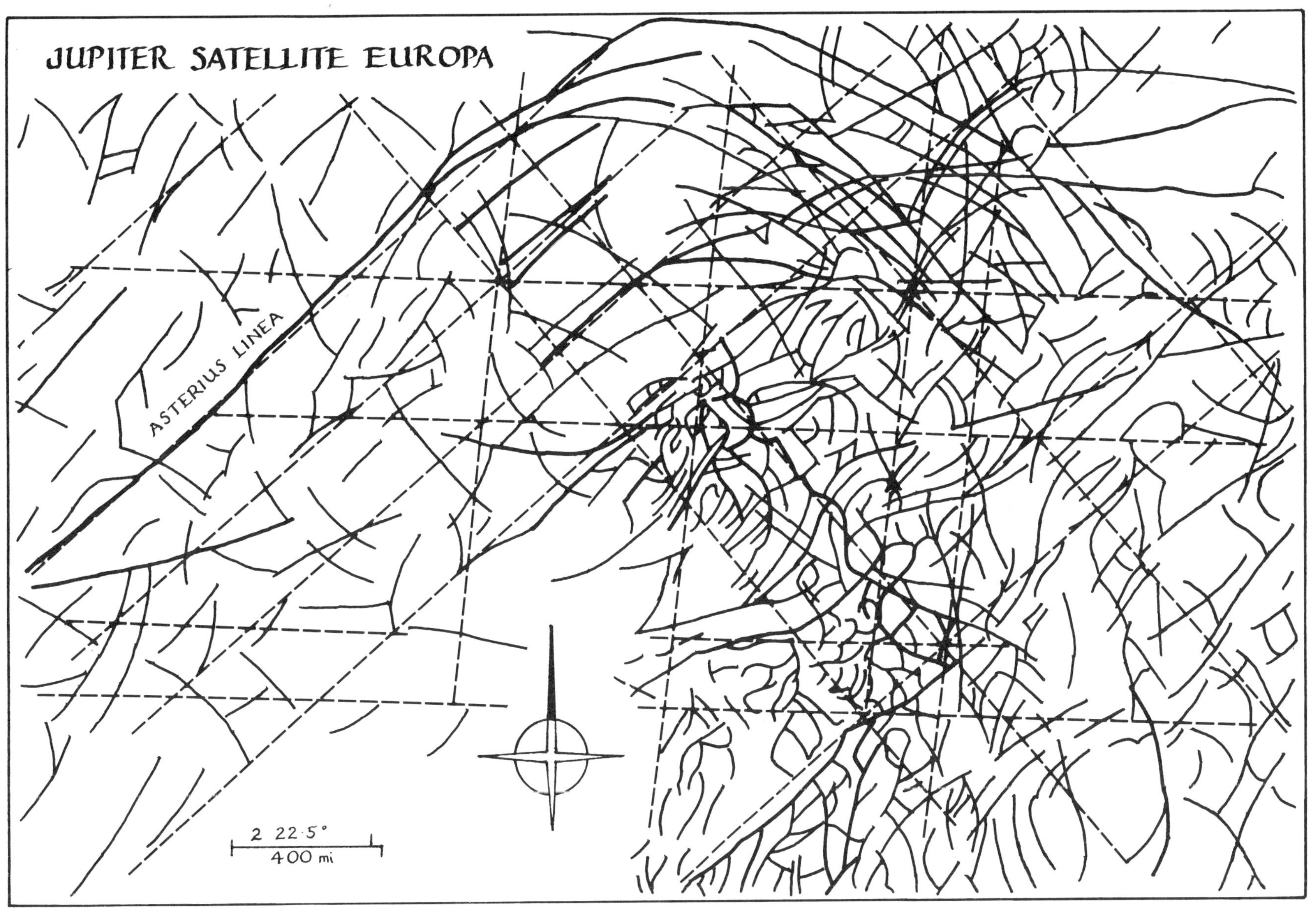
JUPITER SATELLITE EUROPA
ASTERIUS LINEA
2 22.5°
400 mi

5-9 Jupiter Satellite Io

This shows the surface of Io, another one of Jupiter's Galilean satellites, and also another one of those pretty maidens who caught the eye of Zeus and felt the jealous anger of Hera his wife. When Hera found out what was going on, she changed poor Io into a white cow, and then sent a large horse fly, or possibly a cowfly, down from heavenly Olympus to chase her with its bites and stings all over the world. Of course, nothing happened to Zeus who continued circling the sky as Jupiter, along with the essence of Saturn, keeping the universe in order.

Jupiter travels faster than Saturn, completing his circle in 12 years, while Saturn takes 30 years to circle the sky. Each 20 years Jupiter overtakes Saturn and passes him, after Saturn has completed two-thirds of a circle. This is called a conjunction, and during a 60 year period there are three conjunctions that divide the sky circle into three equal parts. At the end of the 60 year triple conjunction period, Jupiter has circled the sky five times and Saturn twice, and they end up together at their starting point in the star pattern. That is to say, almost together, because nature rarely operates with whole numbers, and the Jupiter and Saturn periods are really 11.86 years and 29.46 years, so that a conjunction really happens every 19.86 years. This means that at the end of the triple conjunction period, the two planets come together about 9 deg past their starting point, which represents one-fortieth of a sky circle. This slight difference gave the ancients another very slow hand on the Jupiter/Saturn sky clock that ticked off a full sky circle every 2,400 years, representing 40 triple conjunctions of 60 years each, or 120 single conjunctions of 20 years each. It is also interesting to note that this slow hand moved through one-third of the sky circle in 800 years, representing 40 single conjunctions. What a wonderful sky clock this was, run by the essence of old Father Saturn who handed out measures of time to his son Jupiter each time he passed his father. In addition to the 12 and 30 years of the Jupiter and Saturn orbits, there were 20 years for a single conjunction to mark two-thirds of the sky circle, 60 years for a triple conjunction to mark the full circle, 800 years of 40 conjunctions to mark one-third of the circle, and 2,400 years of 40 triple conjunctions for the slowest hand to go full circle.

But the Jupiter/Saturn sky clock was even more amazing. There is a slight wobble in the axis of the Earth that causes the Sun to shift a bit each year along the sky circle. This is called precession of the equinoxes, and it takes about 26,000 years for the Sun to shift completely around the sky circle, moving through all 12 of the Zodiac figures. A shift through one Zodiac figure takes about 2,200 years, which is very close to the 2,400 years that it takes for the slowest hand of the Jupiter/Saturn sky clock to complete a full sky circle of 40 conjunctions. Over the last 2,000 years there have been 100 Jupiter/Saturn conjunctions that have occurred under the Zodiac sign Pisces, the fish. There remain only 20 more conjunctions over the next 400 years before the slow triple conjunction hand returns to its year one position, and the equinox Sun moves into Aquarius, the water carrier.

This line drawing of Io cannot possibly compare with the shaded and colored NASA map that it was traced from. Io has been described as a giant pizza, with splashes of various shades of red, brown, yellow, and white that cover the surface. Scattered over the multi-colored surface are rounded amoebalike black spots some tens of miles across, that have turned out to be volcanic calderas which are centers of extensive volcanic activity. The calderas are intensely black with surrounding hazy black haloes, and several were erupting during the Voyager fly-bys, sending up sprinklerlike fountains of volcanic material more than 150 miles into the atmosphere. The entire surface of Io is covered with eruptive material that has spilled out and flowed from the calderas. No impact craters can be seen, indicating that the volcanic flows that cover the satellite are relatively recent. It is also probable that the volcanic fluid is thin with a very low viscosity, because the surface is fairly flat with little relief.

Even before the Voyager arrived, both sulfur and sulfur dioxide had been identified on Io. As sulfur cools, it takes on various colors that range from jet black to red and yellow, and sulfur dioxide would be white and frosty on Io's surface. Therefore, it is probable that molten sulfur and its compounds erupt from the calderas, and a sulfurous dust is blown high into the atmosphere that either falls back on the satellite or escapes to supply the vapor glow that trails behind Io as it circles Jupiter. The enormous energy needed for the extensive volcanic activity probably comes from tidal forces acting within Io. Although the distance between Jupiter and Io is about the same as between Earth and the Moon, Jupiter has a thousand times the volume of Earth, and is more massive than all other planets combined. As the distance changes between Jupiter and Io, the satellite is repeatedly squeezed and released, so that the interior of Io remains hot and molten. This tidal action could act as a twisting shear force consistent with the rectangular grain of the surface which trends N40W and N50E. These diagonal directionals can be most clearly seen in the upper center of the figure above the relatively clear Colchis Regio area. A number of the calderas, called Pateras, scattered over the surface also reflect the rectangular grain.

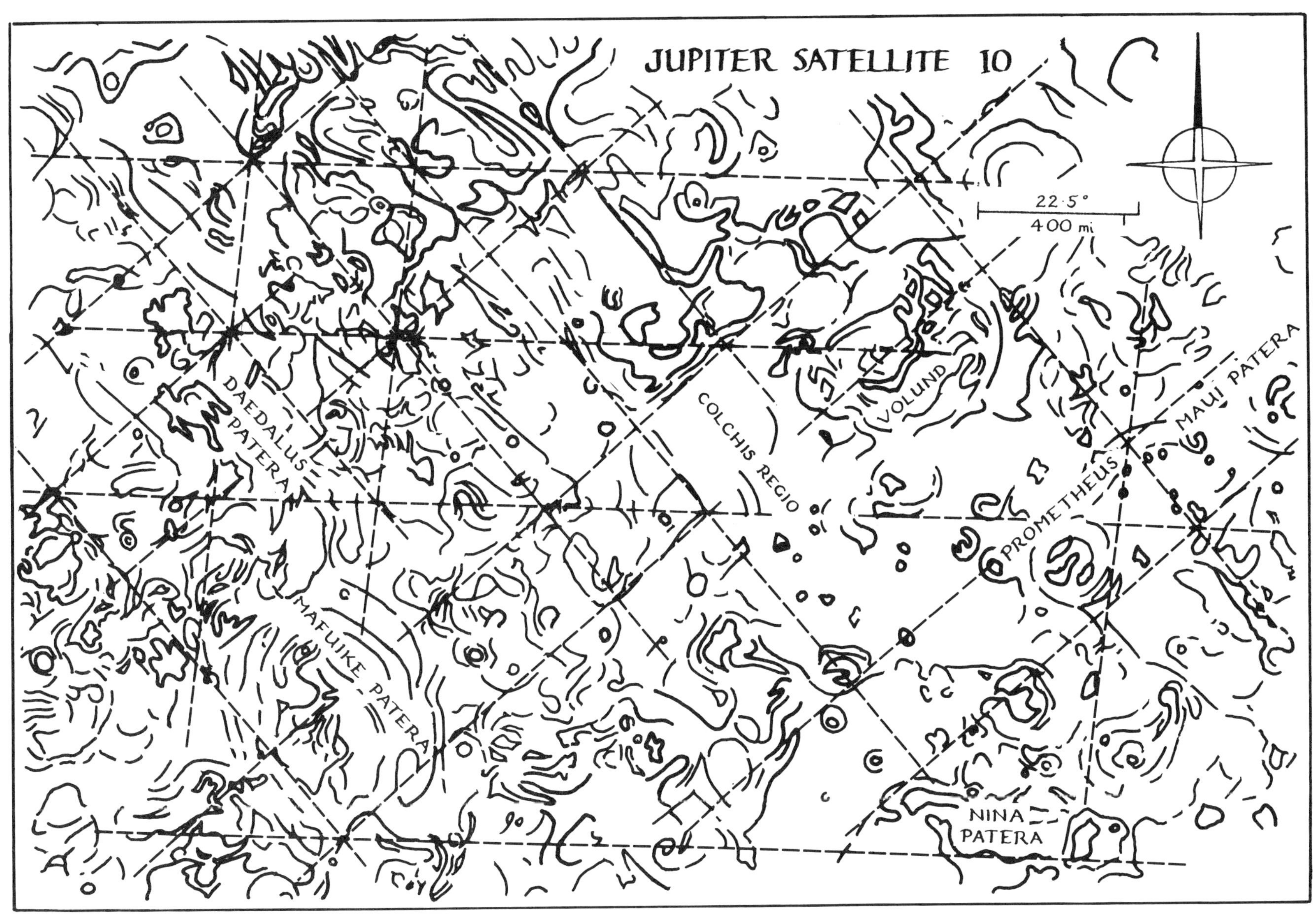

JUPITER SATELLITE 10
22·5°
400 mi
DAEDALUS PATERA
MAFUIKE PATERA
COLCHIS REGIO
VOLUND
PROMETHEUS
MAUI PATERA
NINA PATERA

5-10 Mars Olympus Mons

This shows the huge shield volcano Olympus Mons on the planet Mars. It is the largest volcano known in the Solar System, rising up some 17 miles above the planet's surface and spreading out about 400 miles across its base. Its volume is at least 50 times larger than Mauna Loa, the largest volcano on Earth, which is about 5.5 miles high from a base that spreads out 75 miles across the ocean floor. Although much larger, Olympus Mons has many of the features found on the Hawaiian shield volcanoes. At the summit is a depression called a caldera made up of several smaller circular depressions, and probably formed by collapse when magma below the summit area flowed away. The flanks of Olympus Mons have gentle slopes, typical of the Hawaiian shield volcanoes. However, unlike the Hawaiian counterparts, the upper flanks of Olympus Mons are terraced with concentric stair steps, and there is a high cliff that circles the base of the volcano. Another unusual feature is the aureola, a plain covered with an intricate complex groove pattern that surrounds the volcano out beyond the basal cliff. A portion of the aureola can be seen in the upper left quadrant of the figure.

As mentioned above, there are many volcanic features found on Olympus Mons that are similar to those found on the Hawaiian volcanoes. An eruption from a Hawaiian volcano usually comes from a small vent, and is often periodic with fountaining occurring at fairly regular intervals. When lava is blown up into the air, small particles harden and fall down to pile up as a cone on the leeward side of the vent. The remaining liquid lava slashes down and collects in a lava lake near the vent. Where the lava is exposed to air it hardens rather quickly, forming a crust across the surface. As fresh lava continues to splash down into the lake, it churns up the surface and lava splashes out over the sides where it hardens into a raised ridge around the lake called a levee. Eventually, with the addition of more lava, the lake sur-face rises until the lava breaks out and spills over the levee to flow downhill as a stream.

Lava is thick like molasses and tends to pile up before it begins to flow. This gives the flow front the appearance of a multi-lobed bank with bulges like an amoeba. The lava stream also tends to flow more freely over a steep slope, breaking up into a network of braids that can recombine into a single stream when it reaches a more gentle slope. Similar to the process at the lava lake, a crust forms on the lava stream surface and levees grow along the banks. As the surface crust and levees grow, often they eventually meet and weld together to completely enclose the stream, forming a tunnel called a lava tube. The hard lava crust on top of the stream insulates the molten lava below from the cool air, so that with very little drop in temperature the flow can travel a long distance. Often, a wide lava flow becomes fully crusted over with solid lava, and the hot liquid lava flows below the crust through an intricate and complex network of interconnecting lava tubes of various shapes and sizes. Within this network the flow pattern can change suddenly as tubes become plugged or open up, and the flow is diverted. A lava field formed in this way leaves many distinguishing features. A main channel with prominent levees along the banks can often be seen within the individual flow fingers, and where a lava flow ends, the front is usually lobed. Lava tubes can be traced along a row of pits called skylights where the crustal roof of the tube has partially collapsed. Near the vent where flow begins, hardened lava lakes enclosed within high levees can sometimes be observed. All these lava flow features have been seen on the flanks of Olympus Mons on Mars, indicating by analogy that it was probably formed in much the same way as the Hawaiian volcanoes on Earth. However, caution must always be exercised with conclusions from analogy. Nature is notorious for ending up with similar forms created by completely different processes.

There are many features in the Olympus Mons region that trend along Earth directionals. The summit caldera is elongate, with two small craters at opposite ends which mark a N50E trend. The northwest side of the caldera is linear and also follows N50E, emphasizing the overall trend. Below the caldera on the upper slopes are the stepped terraces, which are especially prominent to the east. The first terrace just below the caldera is a rectangle open to the northwest, with the long sides following two N40W directionals separated by 2.81 deg, and closed on the southeast by a N50E directional. To the east the lower terraces also have sides oriented N40W and N50E, but with the angle of intersection cut off along N7E. On the lower flanks, the grain shows the same characteristics of other radial systems seen on Earth. Groups of linears tend to have a common orientation rather than spreading out like a fan. On the northwest slope the grain trends N40W between the extensions of the long sides of the top terrace, and to the southwest the trend is N50E, corresponding to the caldera trend. North of the caldera the grain trends N7E over a fairly wide area. Although the aureola in the upper left quadrant has no earthly analogy, it reflects the Earth directionals. Two wide linears within the lower right portion of the aureola intersect to form a cross, with the cross bars oriented N40W and N50E. The finer groves just left and above the cross show a N7E grain, while those to the right trend N50E.

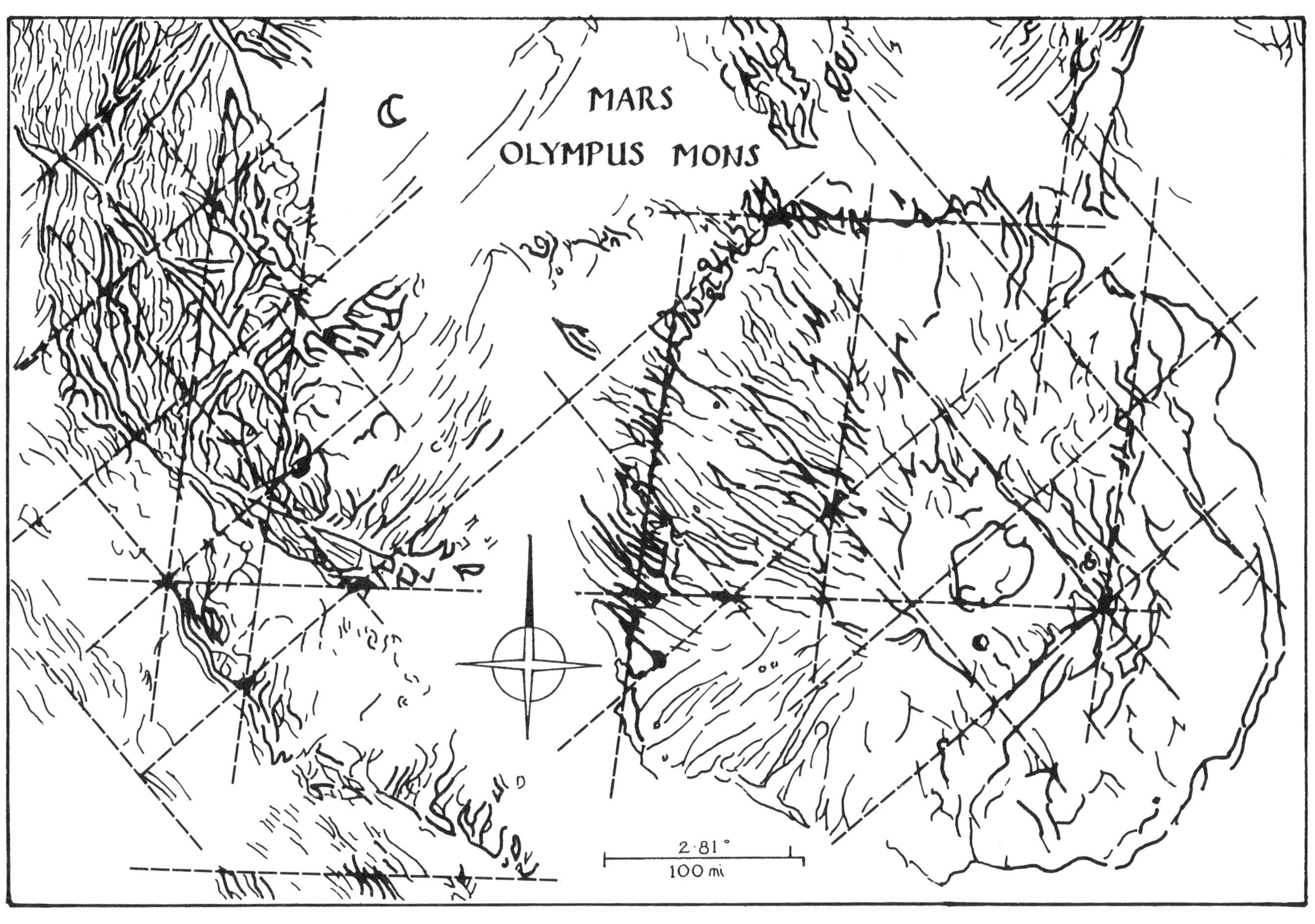
MARS
OLYMPUS MONS
2·81°
100 mi

5-11 Mars Chryse Planitia

Shown here is the southern portion of the Chryse Planitia on Mars that lies just east of the enormous canyon system of Vallis Marineris. The area is crossed by three outflow channels, Simud Vallis, Tiu Vallis, and Ares Vallis that run from the southeast up towards the northwest. A fourth outflow channel called Shalbatana Vallis can be seen on the extreme left running up toward the northeast where it joins Simud Vallis. These large channels are as wide as 50 miles and may run for more than 1,000 miles. They start just south of the lower border of the figure from small chaotic areas that appear disturbed, jumbled, and rough, and end north of the figure on the edge of the Chryse Planitia. Within the wider parts of the channels, a complex intertwining network of smaller channels can often be seen that resembles the braided pattern of a lava stream on a steep slope. On the channel floors there are scared areas where layers of bedrock seem to have been gouged out and stripped off, and other places appear as if they were scoured, leaving parallel patterns of long narrow ridges. Rising above some channel floors are large mesalike rock remnants, that are sometimes shaped like streamlined teardrops and sometimes resemble large rectangular blocks. All these features are much like those found on Earth within the North American Columbia River Basin in the eastern part of Washington State. This area has appropriately been called "the scablands," and it may be that the outflow channels on Mars were formed in much the same way.

About 20 million years ago, a series of volcanic eruptions spread lava over some 200,000 square miles of the northwest United States. Large flows of this type are called flood lavas, with the lava erupting from surface cracks rather than from a volcano vent. The lava flood that created the Columbia River basalts is probably the largest known flow of its kind in the world. The total volume of lava that covered the area is estimated at 35,000 cubic miles or more, and in the Snake River Gorge on the eastern border of Washington State the basalt thickness reaches 4,000 feet. The lava piled up in layer upon layer, which split down along vertical cracks called joints to form groups of packed columns. There is evidence that hills up to 2,000 feet high were buried, while only the tops of higher hills can be seen above the surface. This large lava plateau is tilted down towards the Columbia River on the west which loops around the edge of the flood lavas. On the upraised northern edge of the plateau is the Spokane River which comes off the mountains from the east and flows west to join the Columbia.

During the most recent ice age that ended 10,000 to 15,000 years ago, the North American continental glacier front followed the Spokane River, leaving the basalt Columbia Plateau to the south free of ice. During the ice age years, clouds of fine silt were blown in by the wind and dropped on the plateau, forming a blanket of soil up to 100 feet thick which was piled up in low rolling hills that looked like a stormy sea to some people. Towards the end of the ice age the continental glacier began to melt as the climate warmed. However, just east of the Columbia Plateau high in the mountains, lobes of ice remained around the headwaters of the Spokane River. These lobes formed an ice dam that held back water east of the mountains, creating the great glacial Lake Missoula that covered almost 3,000 square miles and was 2,000 feet deep in places. But the ice dam was melting, and it was just a matter of time before it ruptured and suddenly released the Lake Missoula waters. The flood that followed may have been the greatest fresh water flood that ever occurred on Earth, with a peak flow equal to 100 Mississippi Rivers at flood stage. The huge bank of water first roared down the Spokane River just south of the glacier front, but almost immediately overflowed the river banks to surge southwest across the Columbia Plateau and eventually pour into the Colombia River. The flood waters may have been as deep as 800 feet, and they ran for possibly two weeks. The huge flood of rushing water scoured the plateau, stripping off over 2,000 square miles of the soil blanket and ripping out strips of basalt to form channels up to 400 feet deep. Resistant remnants of rock on the channel floors were sculptured into teardrop islands or rectangular mesalike blocks. The entire area was transformed into scablands that in many ways matched the Mars outflow channel areas.

It has been suggested that floods of melt water may have somehow pushed up through the Mars surface and poured out of the chaotic areas to rush down the channels and shape the scabland features on Mars. Explanations on just how this may have happened are largely speculative due to the sparse information available. A better understanding will require more information, and that can only come from another Mars space mission. The grain of the Chryse Planitia area runs N40W, but is crossed by a number of linears running N50E. As with so many examples, this gives a rectangular aspect to the pattern. This is especially clear on the west where the N50E Shalbatana Vallis joins the N40W Simud Vallis in a right angle. Several linear features reflect the N7E directional, as reflected along the east side of the large block in Simud Vallis at the bottom center of the figure.

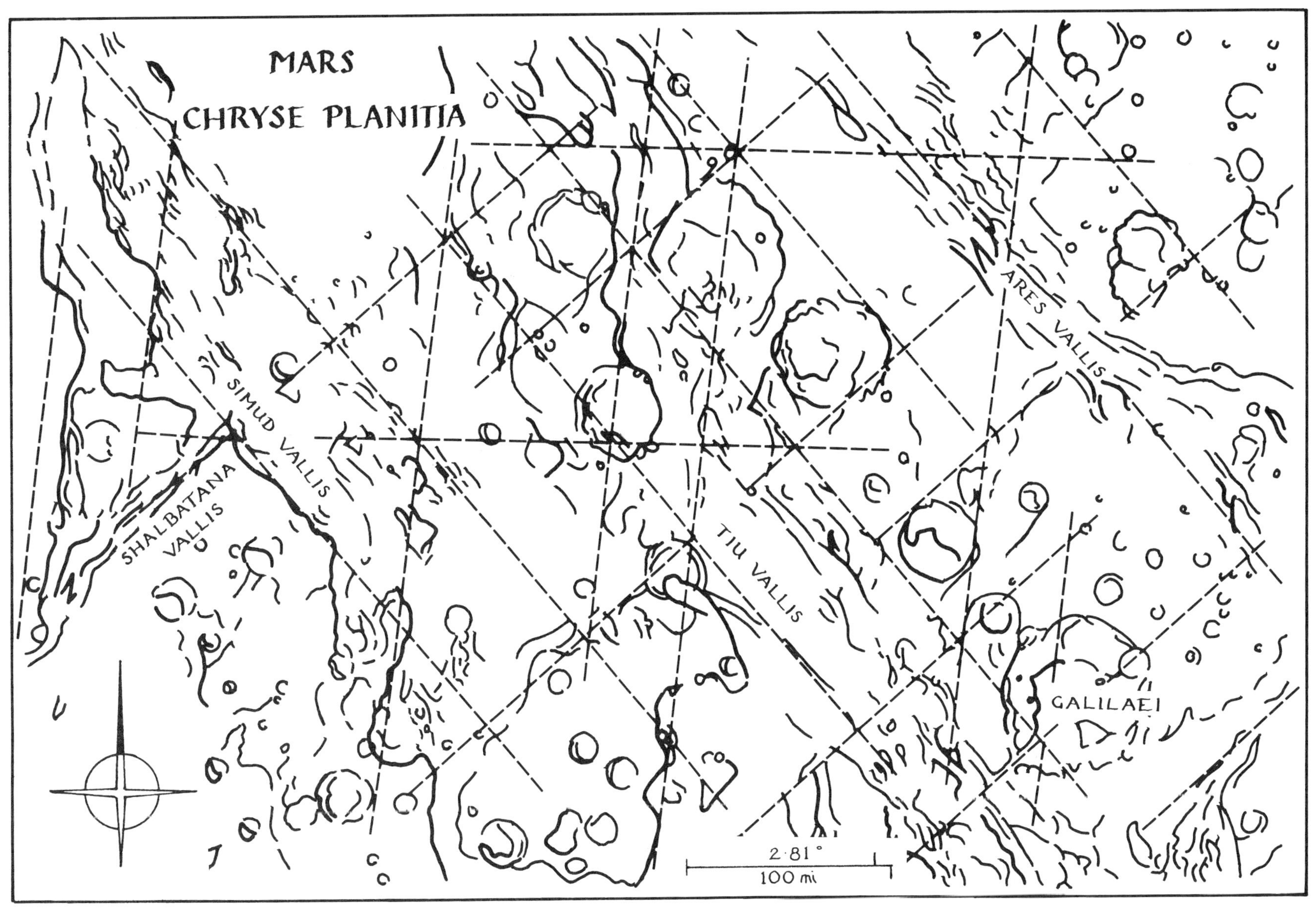

MARS
CHRYSE PLANITIA
SHALBATANA VALLIS
SIMUD VALLIS
TIU VALLIS
ARES VALLIS
GALILAEI
2·81°
100 mi

5-12 *Moon Lava Flows*

Shown here are the dark plains on the near side of the Moon that faces Earth. The outlines of the plains were traced from a Mercator Projection map in Antoin Ruki's book, "Moon, Mars, and Venus." At full moon imagination can turn the inkblot pattern of the patches into many wonderful pictures that various cultures have called lizard, turtle, hare, or man in the moon. In Hawaii, the patches became Hina, the woman in the moon who spends her time beating out bark to make tapa cloth. On Earth, she grew sick and tired of being ordered around by her husband and cleaning up after her children, so one night she jumped up into the Moon, leaving all her troubles behind. When her husband saw her jump, he was just able to grab one of her legs which twisted off and was left behind as she escaped. It is said that Hina's home is under the sea, and in midwinter when the full Moon is farthest north along the horizon, she is certainly a beautiful sight rising up out of the ocean at sunset.

In all cultures the Moon has always been thought to influence things on Earth. It is said that the full Moon brings more babies, but also more lunatics. Crops are planted in rhythm with the Moon's phases, and the Jewish and Islamic religions still use calendars in tune with the Moon. And of course, there is the influence of the Moon on the ocean tides. It is natural to think of the Moon as circling the Earth, but that is not quite right. The bodies swing around each other like two weights at the ends of a twirling rope. The bola used by the South American cowboys works much the same way. As the Earth and Moon twirl around each other, both bodies tend to stretch out slightly like a lemon drop as they are pulled inward toward each other by gravity and pulled outward by centrifugal force. This causes the oceans on Earth to bulge up a bit, both on the side facing the Moon and on the opposite side. As the Earth rotates, these two bulges travel around the Earth as broad ocean waves, causing high tides twice a day in many parts of the world. Of course, there are the continent and island barriers that block or divert the tide waves, and also difference in water depth, all of which contribute to the great diversity in tidal patterns around the Earth. Twice a month at full and new Moon, the Earth, Moon, and Sun line up in what is called "syzygy," pronounced "sizz-a-gee." When this occurs, gravity forces from both the Sun and Moon reinforce each other, and high spring tides develop. As used here, the word spring has nothing to do with "springtime." Every year or two, gravity is further reinforced when the Moon swings closer to Earth in perigee, and even higher tides called "proxigean spring tides" occur. If high winds come up at the same time as the higher tides, there can be extensive destruction along the shore. Three extremely high proxigean spring tides are due on December 2, 1990, January 19, 1992, and March 8, 1993. Hopefully, the sea will be relatively calm when this happens.

When stargazers first saw the Moon through telescopes it was naturally thought that the dark patches on the surface were bodies of water, so they were called "mare," which means "sea" in Latin. The bulges on the boundaries of the larger areas were thought to be "bays" and were called "sinus," again in Latin. At that time it was believed that good weather came with the waxing Moon as it grew larger, so that areas seen during the first two quarters were given good weather names such as Mare Serenitatis, serene sea or Mare Tranquillitatis, tranquil sea. Bad weather came with the waning Moon as it became smaller, so that areas seen during the last two quarters received names like Mare Nubium, cloudy sea, Mare Imbrium, rainy sea, or Oceanus Procellarm, stormy ocean. Far to the north was Mare Figoris, frigid sea, corresponding to the Arctic Ocean on Earth. However, when man went to the Moon and brought back rock samples, it was found that the dark plains were relatively smooth areas of volcanic basalt, much like the Columbia Plateau basalts on Earth, that formed from large flood lava flows that erupted on the Moon some three to four billion years ago and collected in the lower areas, such as the large multiringed impact basins.

As with many other examples, the edges of the dark lava flows generally trend N40W and N50E, giving the effect of a rectangular or diamond pattern. It will be remembered that in the strain analogy these directionals represent the diagonals of the strain rhombus that intersect in right angles and follow peak lines of deformation with N40W shortened and N50E lengthened. On the left, both Oceanus Procellarum and Mare Imbrium are enclosed within large rectangles that overlap, forming an arrow pointing west. The long sides of these rectangles are separated by 45 deg hierarchical intervals. North of Oceanus Procellarum is the bay Sinus Poris with its west edge following a N7E directional, and north edge a W2N directional that also traces the southern boundary of Mare Figoris. The dark areas right of center seem to follow a stair step pattern with a N40W trend for Mare Serenitatis that shifts to N50E along Mare Tranquillitatis, and then back to N40W along Mare Nectaris. In the lower right corner is a group of smaller Maria that are boxed in by rectangles, again formed of N40W and N50E directionals. The overall correspondence of the lava flows on the Moon and the Earth directionals is quite striking.

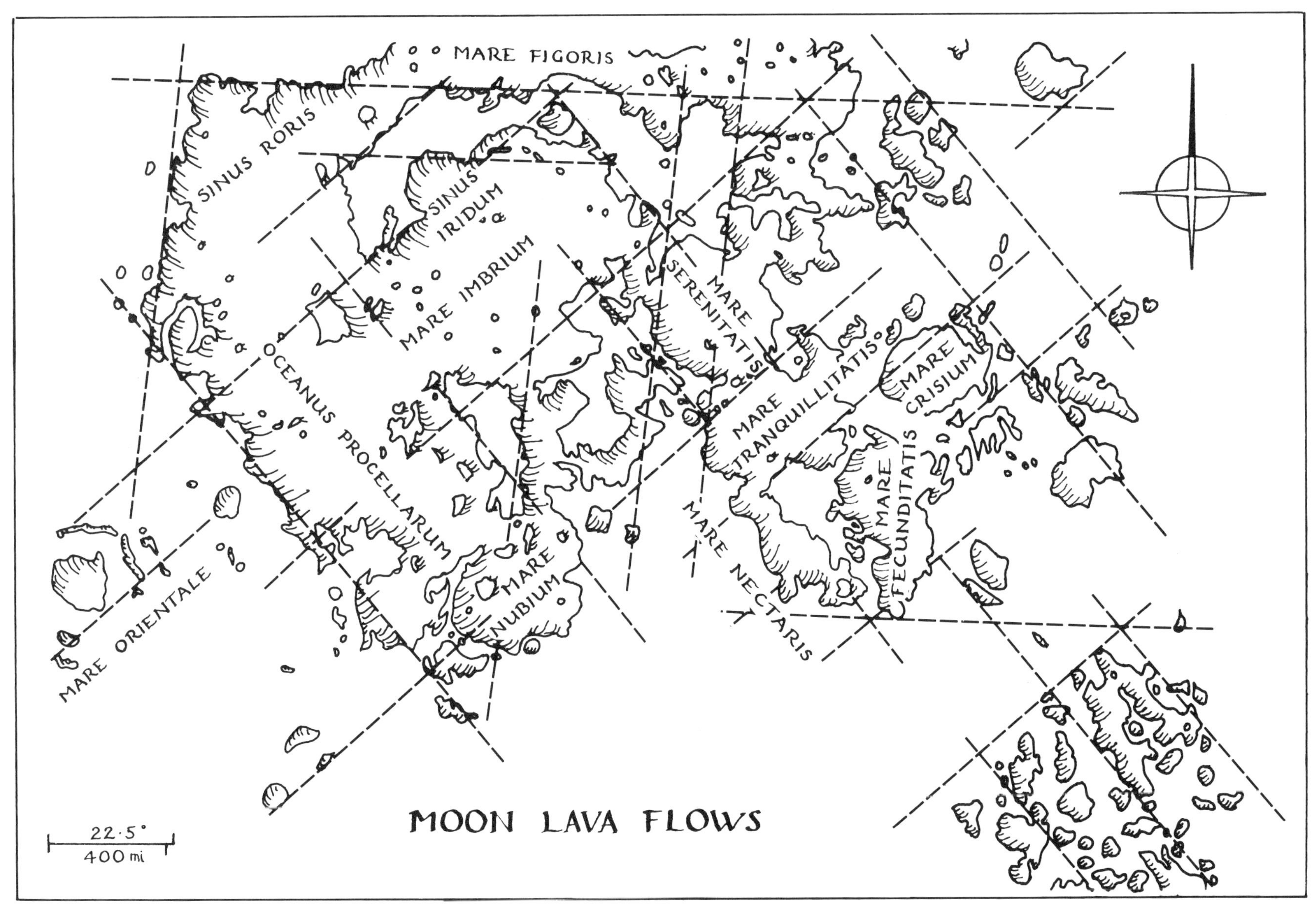

MARE FIGORIS
SINUS RORIS
SINUS IRIDUM
MARE IMBRIUM
MARE SERENITATIS
OCEANUS PROCELLARUM
MARE TRANQUILLITATIS
MARE CRISIUM
MARE FECUNDITATIS
MARE NUBIUM
MARE NECTARIS
MARE ORIENTALE
22.5°
400 mi
MOON LAVA FLOWS

In this chapter surfaces of several Solar Systems bodies beyond Earth were examined. A number of features were shown that included impact craters, fracture systems, volcanic features, and flow channels. It was found that all four of the Earth directionals appear on the Solar System bodies with varying degrees of clarity. The most striking pattern was found in the probable fracture system on the surface of Jupiter's ice covered Europa, where the distinct black lines that crisscross the satellite correspond closely with the Earth directional pattern. The lava flows on the Moon and the outflow channels on Mars also show an excellent fit with the directionals. Other examples were somewhat less clear, but all reflected the Earth directional pattern within the surface features. The degree of clarity may well be related in part to the resolution of the photographs, many of which were taken at a considerable distance on fly-by missions.

The appearance of the Earth directionals on other bodies raises rather formidable questions. It is probable that all bodies in the Solar System experience some torsional force around the axis. No Solar System body is perfectly round, nor is the material in the body perfectly distributed, and all spin about an axis that is tilted in varying degrees. Therefore, it might be expected that the gravity and acceleration forces acting on the Solar System bodies would be applied unsymmetrically. This would bring a torsional or shear stress to bear, twisting the body about its axis and stretching the surface into a strain rhombus pattern. However, there is no reason to expect that the amount of deformation, or strain, would be the same in all cases. The amount of strain from a shear stress is measured by the angle of twist, and this is dependent on both the amount of force applied and the make-up of the body being twisted. The greater the force and the greater the elasticity of the body, the greater the resulting angle of twist. For example, if the same twisting force were applied, a bowling ball would deform only a little, but a balloon would deform a lot. The Solar System bodies are widely

different in size, composition, density, structure, and environment, and have interiors that range from liquid to solid, giving them highly variable elasticities. In addition, the axis rotation, orbit revolution, and distance to neighboring bodies differ greatly, and this could effect the amount of shear force applied. All in all, a rhombus strain pattern with four prominent lines might be expected, but it also would be expected that the directional orientations would be unique for each body. Contrary to expectations, the examples showed the strain pattern on all bodies with the directional orientations identical to those observed on Earth.

The presence of the same directional pattern on all Solar System bodies requires some common force acting throughout the Solar System. It seems probable that a common force of this nature must be tied in some way to the Sun which dominates the entire Solar System. Although there are many known forces that come from the Sun, none seem to be directly tied to the directional pattern. However, as mentioned before, it is an interesting coincidence that the angle of twist in the strain analogy corresponding to the N7E directional is identical to the 7.2 deg tilt of the Sun's axis as related to the orbital plane of the planets. It may be that the force in question is in some way connected with the rotation of the Sun about its tilted axis. One known force that is directly tied to the Sun's rotation is the solar wind which is thrown out from the Sun as it rotates. As this plasma wind speeds away from the Sun, it takes the form of a giant pinwheel, symmetrical with the 7.2 deg tilted axis of the Sun. Perhaps this tilted force field sweeping out across the Solar System reacts in some way with the individual bodies in the system. But this is merely unrestrained speculation and explains nothing. For the present, there are only the isolated observations that must be set aside to wait hopefully for something more concrete.

The voyage taken across the Solar System in this chapter was called a diversion, with an entirely different class of material brought up for discussion.

However, the main topic of the book is the Earth and the directional pattern found within the network of surface linears. The results of this chapter's discussions illustrate both the benefits and problems associated with diversions. It is not unusual that similar patterns can be found in different sets of examples, and that new avenues of approach are opened. However, the discovery almost always raises a multitude of new questions that often serve to fragment the direction of analysis. A major problem of diversions in geologic analysis is when to stop work on the sideline and get back to the main topic. It is very easy to continue indefinitely on the diversion as new questions arise, to the detriment of the main subject of analysis. Diversions can be fascinating, challenging, and fun, and the continuum of nature provides an unlimited supply of questions branching off from the main topic. In analysis it takes a conscious, decisive effort to stop peripheral work and move attention back to the main stream, leaving behind many unanswered questions. Therefore, attention will be directed back to Earth in the last two chapters. The possible usefulness of the directional concept will first be examined, after which the discussion will center on several of the major questions that remain unanswered.

6 THE APPLICATION

USING THE PATTERN

A basic requirement of all geologic analysis is usefulness. A study may be interesting, imaginative, and well written, but if it does not eventually prove useful, it will sooner or later fade into obscurity. However, usefulness is a very relative concept and depends on many things, including the overall cultural environment and the personality and interest of both author and audience. The public has always been eager to accept wild speculation if it relates to some basic cultural belief. Some studies that attempt to provide a scientific basis for religious writings might fall into this category, and they can be extremely useful to the public as a rationalization of existing beliefs. Many geologic studies are written largely to support an accepted concept, and they also can be very useful to members of the profession in bolstering credibility. Papers written for graduate degrees and articles in professional journals are sometimes of this nature. Any well written and logically consistent study that reinforces and confirms an accepted belief can be considered useful.

There are many geologic analyses that stop short of explanation and concentrate on the collection, sorting, and organization of information. This type of study is often accomplished by those rare individuals who enjoy accumulating diverse observations and sorting them into some organized system. This personality is generally not primarily interested in the pattern of interconnections between categories. The studies can be extremely useful, because they provide the building blocks for all other analyses. Lifetimes have been spent collecting, describing, and classifying the physical properties of rocks, while others have worked long hours in the laboratory determining chemical properties of minerals and organizing the results into a useful system. Today there are many sophisticated data collection methods that spew out enormous quantities of information, and a large proportion of geologic analysis time is routinely and necessarily spent in reducing this jumbled mass into a useable organized pattern for study. The value of these studies is often underrated by those who most use the results in their own work. The development of geologic explanation is essentially the search for generalized interconnections within the catalogue of classified observations. Those who compile the catalogue point the way for others who use the material for explanation.

There are other personalities who dread the boring routine of collection and classifying, but thrive on puzzles. Their interest is to find generalized relationships that will bind together seemingly unrelated categories of observations. Not too long ago this took the form of long and detailed written arguments. More recently however, collectors have refined the organizational classifications, moving toward quantified statistical measures. This has provided the explainers with the numbers needed for mathematical analysis, and eventually there has evolved the sophisticated mathematical models that dominate the geologic analysis of today.

Whatever the method, a geologic explanation can be useful in many ways. It is first of all interesting and satisfying because it provides a rationale for the processes that occur around us in the natural world. People have always searched for reason in nature, but often were forced to fall back on the supernatural for lack of something better. A geologic explanation helps to shift this dependence from the unrational supernatural to the rational natural. Often the explanation provides the basis for tracing geologic processes backward in time to give a reasonable and consistent geologic history. Predictions follow, with the extrapolation of the history forward in time. Explanation is also used in estimating an existing but unseen configuration that lies below the surface, which is merely another type of prediction unrelated to time. Of course, the usefulness of any prediction is highly dependent on the uncertainties surrounding both the explanation and the available data on which it was built.

For the general public, the word "usefulness" has a more restricted meaning. It deals with the usefulness of a concept within the practical world, or how the concept might be used for the economic or social benefit of people. This restricted meaning forms the basis of this chapter, where several examples will be given that point to possible practical uses of the directional pattern. The examples are merely illustrative, and no attempt has been made to develop conclusions or recommendations. Use of the pattern in a specific geologic analysis would require a far more detailed study of the many diverse aspects that apply to a particular case. The main focus is the use of the directional pattern as an organizing tool to provide an outline or framework within which surface features can be fitted. Extrapolations or interpolations within the pattern might well be improved, providing a more substantial base for predictions. An analysis is never a routine process, and the analyst needs all the help he can get. The directional pattern could prove to be another useful analysis tool supplementing those that now exist.

The illustrations in this chapter were traced from various professional journals found in many corners of the world over a considerable period of time. A number of the sources were rather carelessly not recorded because no book was planned at the time. This is unitentional and a sincere apology is offered to those where credit is due. As mentioned before, geologic analysis rests on the work of many who have spent a lifetime collecting and organizing observations.

6-1 Oregon Clackamas River

This shows the shifting channel of the Clackamas River, located in the northwest United States, over a period of a little more than 50 years. The Clackamas begins in the Cascade Range between the volcanic peaks of Mount Hood and Mount Jefferson. It flows west into the Willamette River, that in turn empties into the Columbia River at Portland, Oregon. The name "Clackamas" comes from an indian tribe that lived along its banks and spoke a dialect of the widely distributed Chinookan language that formed the basis of the Oregon trade language, called the Chinook Jargon, which was used during the early settlement period. Located at the juncture of the Clackamas and Willamette Rivers was Oregon City, the end of the Oregon Trail, and the two river valleys were major settlement areas for the many emigrants who traveled that road.

The age of a stream generally refers to its flow characteristics rather than its life cycle. It is called "young" upstream near its source in the mountains where it is relatively small and fast flowing as it tumbles down the steep slopes. It is here that the rapidly moving water cuts into the ground and gouges out rocks and soil which it carries down into the lowlands. As the water continues to erode the slopes, valleys are cut out that gradually widen and deepen as more material is washed away. Downstream in the relatively flat lowlands where the stream nears its outlet mouth it is called "old." Here the stream is broad and slow moving as it meanders back and forth across its flood plain, depositing its load of rocks and soil that was picked up in its young upstream portion. Throughout its course from source to mouth the stream is never truly straight, but tends to follow the path of least resistance as it flows downhill, changing its course with the lay of the land. Any shift in direction is reinforced by the eroding force of the rushing water. As it swings around a bend, it continually washes away the outside bank and drops sediment on the inside, making the curve

slowly grow larger. Of course, the curve cannot continue to grow indefinitely, and eventually sediment will build up to dam the channel and shift the flow direction. This process operates in all streams, but is least effective in the younger upstream stage where flow is relatively fast down the steeper slopes. It is most effective in the older stages of the stream on the flat lowlands where material from the upper slopes drops out and is deposited in the flood plain meanders.

The portion of the Clackamas River shown in the figure might best be classed as "middle aged." In its younger stage on the Cascade slopes it would tend to be somewhat straighter and less prone to shifting its course. In an older stage it would tend to show many intricate loops as it meandered back and forth across its flood plain. The Clackamas flows into the Willamette only about 15 miles downstream, northwest of the upper edge of the figure, so that it never really reaches an old stage.

The figure shows four different courses of the Clackamas over a 56 year period. These courses are marked by solid lines alternating with closely spaced dots. The earliest course in 1914 uses three interspaced dots, the next 26 years later in 1940 uses two dots, the third 15 years later in 1955 uses only one dot, and the most recent course in 1970 is traced with solid lines and no dots. The river enters the figure at the bottom right along a N40W directional that can be extended up towards the northwest to again connect with the river just before it crosses the upper edge and leaves the figure. Between the upper and lower extremes of the course that lie on this base N40W directional, the river loops up to the north and then back around to the southwest. Starting from the bottom of the figure, the stream in the eastern part of the loop has changed course many times within a relatively broad strip that is defined by two N7E directionals separated by a 19.8 sec hierarchical interval. Within this strip there is an intricate maze of old stream channels that reflect the N40W and N50E diagonal directionals. The most

recent 1970 channel continues north a short distance, then loops around to the southwest along a N50E directional until it meets the base N40W where it abruptly shifts direction and runs up and out of the figure. The 1955 river course cuts across the loop along a N40W directional just 19.8 sec east of the base N40W directional, and 39.6 sec west of another N40W directional that just touches the top of the loop. These three N40W directionals are part of a series of five spaced 19.8 sec apart, and running from the top of the loop down to the bottom of the figure where the 1970 channel enters.

The entire shifting configuration of the river can be organized within the directional pattern, providing a framework for historical channel changes. This organization of diverse information in a systematic way simplifies visualization, and provides an understandable base for analysis, review, and presentation. As work proceeds, progress can be reported within a standard format to the benefit of all concerned. This allows for the entire process of analysis to proceed more logically, from the initial sorting of data to the presentation of final recommendations.

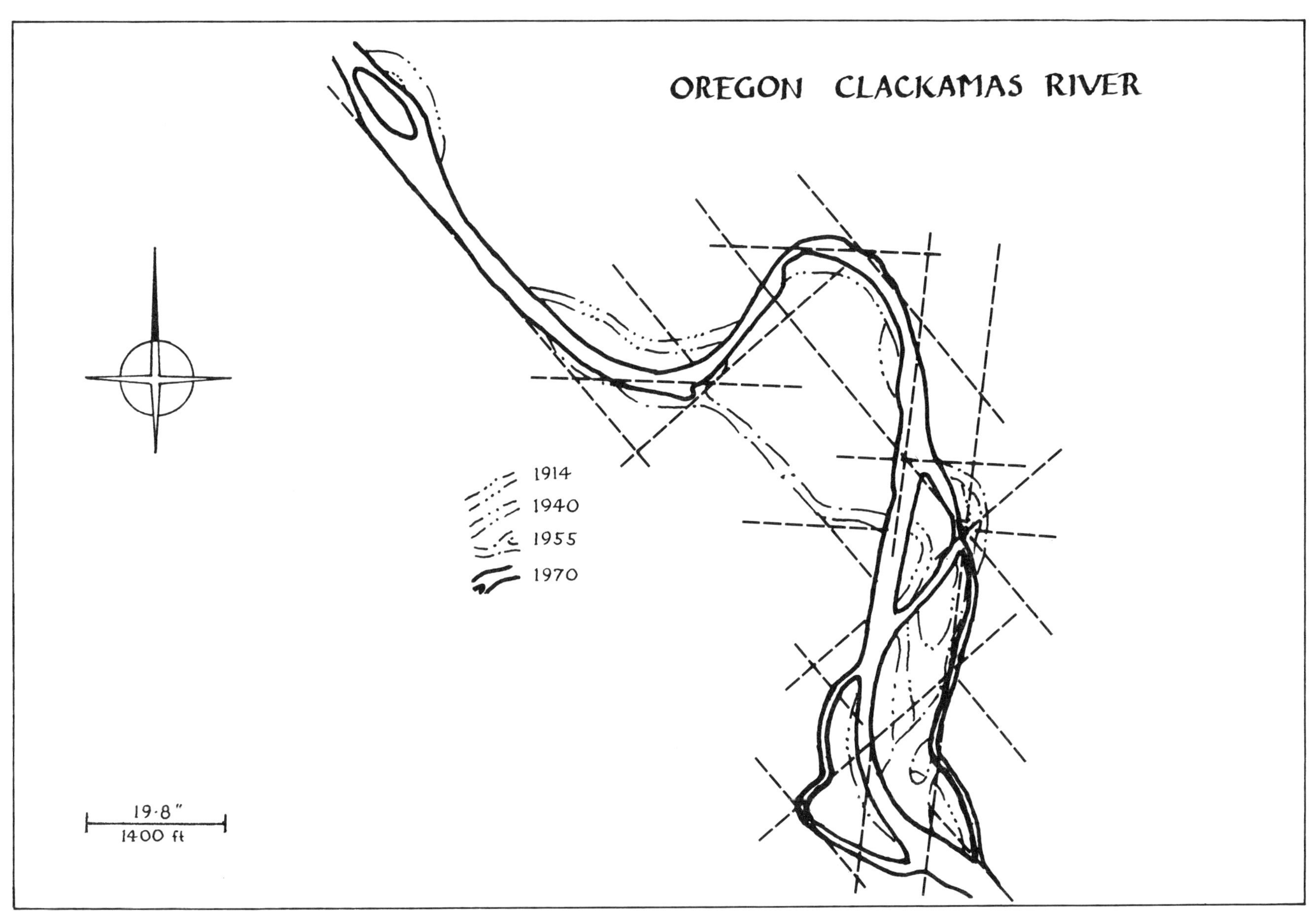

OREGON CLACKAMAS RIVER
1914
1940
1955
1970
19·8"
1400 ft

6-2 *Mississippi Meanders*

This shows a portion of the Lower Mississippi River near the small town of Arkansas City that lies in the southeast corner of Arkansas State. The Mississippi begins near the border of the United States and Canada, flows south as it cuts through the middle of North America, and finally empties into the Gulf of Mexico. About halfway down its course through the United States it is joined by two large rivers. On the west the Missouri River comes off the Rocky Mountains and flows southeast across the dry western plains to join the Mississippi at St. Louis. Just south of St. Louis is Cairo at the southern tip of Illinois State, where the Ohio River enters the Mississippi after leaving the Appalachian Mountains to the east and flowing southwest across the agricultural belt of the United States. With the influx of these two large rivers, the Mississippi enters its old age and is called the Lower Mississippi, which runs from Cairo to the Gulf of Mexico.

The bank of an old age stream has been eroded until it is essentially in equilibrium, so that mud and sediment are merely shifted from place to place. The stream moves down a gentle slope and loops back and forth across a flat wide valley called a "flood plain." As sediment is shifted, the loops, called "meanders," also change and shift within the flood plain, and natural levees bordering the old stream banks outline the dry loops called "meander scars." If water remains in these old meanders they are called "cutoffs" or "oxbow lakes," such as the Beulah and Caula Point Lakes shown in the figure. The broad flood plain valley through which the river wanders is nature's way of taking care of the excess water that the river channel cannot handle. If the river flow becomes so great that the banks overflow, the water spreads out and is contained within the flood plain. This unpredictable flooding happens often and is a natural characteristic of old age rivers.

The Lower Mississippi flood plain is some 50 miles or more wide and stretches for about 600 miles from Cairo to the Gulf. It is a tempting settlement area with fertile black soil on level land adjacent to a river, so that transportation is convenient and water readily available. Just as people moved into the beautiful earthquake-prone fault valleys in California, they also moved onto the rich bottom land of the flood-prone Mississippi flood plain. The water carried by the Lower Mississippi is drained from a very large area that stretches from the snowy Rockies to the wet Appalachians. Winter rains fall on the southcentral part of the area, and early spring rains, along with the snow melt, drain from the north and east. Then somewhat later in early summer, the snow melt from the high Rockies to the west flows into the river. Timing is critical in this sequence, so that late spring rains or an early snow melt in the Rockies can drastically increase the water draining into the Lower Mississippi. This happened in early 1927 when 26,000 square miles were flooded and the river rose to a record 57 feet at Cairo. Damage was estimated at about one billion present day dollars, and more than 200 people lost their lives. Since then the government has worked on a flood control program, and more than 2,000 miles of levees and dikes have been built to contain the river, or allow it to overflow into designated areas.

However, people continue to pour into the flood plain, bringing with them more roads, shopping malls, and parking lots. This increases the flood hazard because of the greater number of people and structures at risk, but it also tends to increase the flood height due to the smaller area available for the spreading flood waters. There is a second factor that increases the flood hazard, and that is the river's doing. An old age river continually drops sediment and raises its own bed, and in its natural state would eventually overflow and loop out in a different direction. However, if the river is trapped in its course by levees and cannot break out and shift to a new direction, sediment keeps dropping, the river bed keeps rising, and the levees must be built higher and higher. It represents an attempt by man to force nature into an unnatural process within an unnatural pattern, and the consequences were felt in 1973 when the "flood controlled" Lower Mississippi flooded and St. Louis recorded the highest flood level in its history. Damage was estimated at one billion dollars, about the same as the 1927 flood. since then, the levees continue to be built higher as the river bed rises, and the conflict with nature continues.

An effective practical solution to a problem that involves nature often requires a design that complements the natural environmental processes. The first step is the development of an organizational pattern that reflects the natural world. As shown in the figure, the meander grain tends to follow the directional pattern primarily within the N40W and N50E diagonal directionals, which could indicate preferred flow directions along those directionals. It may be that the changing river course could be analysed within the directional framework, leading to a more fundamental understanding of the natural process and a better idea of the consequences of flood control alternatives. A program might then follow based on a cooperative effort between man and nature instead of conflict.

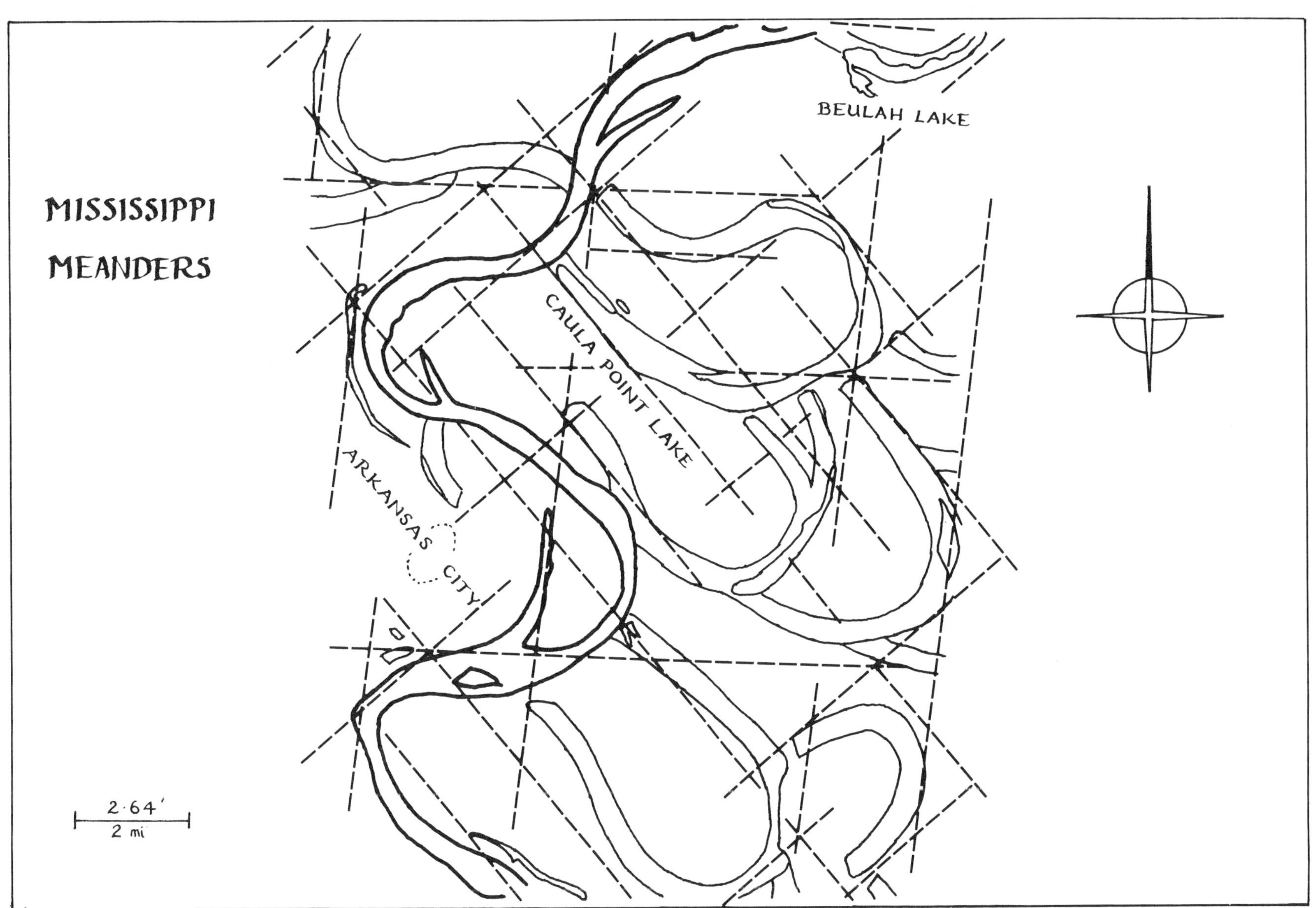

MISSISSIPPI
MEANDERS
BEULAH LAKE
CAULA POINT LAKE
ARKANSAS CITY
2·64'
2 mi

6-3 Iran Oil Fields

This shows the major oil fields of Iran, part of the Middle East oil field complex that contains at least 65% of the world's known oil reserves and is grouped around the Persian Gulf. These Iranian fields, lying just north of the Gulf, hold some 10% of the world's reserves, about the same as all the reserves in North America or Russia. Iran also has enormous natural gas reserves estimated to be exceeded only by those in Russia. West of the Iranian fields are the Tigris and Euphrates valleys in Iraq, and to the northeast is the 12 to 14,000 feet high Zagros Mountain Range that lies within the Zagros Fault System shown earlier.

An oil field outline represents the limits of an underground oil reservoir, which is completely different from a lake or water reservoir above the ground. An oil reservoir is a large body of rock containing oil in the tiny spaces, called "pores," that exist between solid grains of rock. A rock is generally not completely solid, but rather is made up of a mass of small pieces welded together that appears to be a hard chunk of material. However, within the solid-looking mass is an intricate maze of microscopic pores that are often interconnected, so that a fluid can flow through the rock. Deep underground, these pores usually contain water under high pressure, but in an oil reservoir, oil has flowed into the porous rock and pushed out most of the water, leaving oil filled pores that still retain a small film of water clinging to the grains of rock. The porous rock that holds the oil in the Iranian reservoirs is called the "Asmari Formation," and is made of the same limestone that is often used for building blocks when found at the surface.

But an oil reservoir is more than just porous rock. When oil flows into the reservoir it must somehow be held or trapped there so it does not escape and seep up to the surface. If water and oil are mixed, the oil tends to bubble up through the water and float on the surface. Much the same thing happens underground, where oil bubbles up through water filled rock pores until it reaches a sealing layer of solid non-porous rock where it tends to collect underneath the rock seal. In the Iran fields, the sealing non-porous rock lying just above the porous Asmari limestone is called the "Gachsaran Formation." Both these formations are very thick, with the Asmari reservoir rock up to 1,500 feet thick and the sealing Gachsaran rock above from 2,000 to 4,000 feet thick.

In addition to the porous reservoir rock below and non-porous sealing rock above, the reservoir must also be shaped to hold the oil, just as a pond must be shaped to hold water above the ground. However, oil underground tends to flow upward, so that the pond shape must be turned upside-down like a hat or umbrella to trap the oil underneath. In Iran the oil reservoir traps were formed by enormous lateral forces that wrinkled up the Earth's crust into a series of huge folds, with the top of the fold as much as five miles above the bottom of an adjacent fold. As shown in the figure, oil is trapped in many of these folds, forming some of the largest oil reservoirs in the world with the oil filled rock sometimes more than a mile thick and stretching out laterally more than 25 miles in some areas. These oil fields are not only gigantic, but also contain oil wells that produce oil at extremely high rates. When rock is folded it tends to crack up, forming a fracture maze, especially along the top of the fold where the oil is trapped. Many oil wells are drilled into these fractures which act as pipes or channels that can collect oil from the reservoir and carry it rapidly into the well.

In an oil exploration program the geologic objective is to find porous rock that fluid can easily flow through, with a non-porous layer of sealing rock above, and structurally shaped to hold oil. The analysis that determines where a well might be drilled is usually based on the extension of information from known areas out into the unknown. The information can come from many sources that include rock samples from both the surface and underground from drilled wells. Often small explosions are used to generate earthquake waves that reflect back from certain rock layers to indicate the depth of the rock. An exploration geologist attempts to put these known scraps of information on maps, and then use them to predict what the unknown areas might look like. This requires arranging the information into some pattern that can be logically extended into areas where information is sparse or lacking.

The directional pattern often fits an oil field pattern, and might provide in some cases a logical framework for prediction. For the Iranian oil fields, the major trend is N40W, consistent with the predominate trend of the Persian Gulf. In many cases, Individual oil fields are separated from neighboring fields by a directional from the hierarchical pattern, giving an overall impression of fields hung from the directional framework. The correspondence between the oil field pattern and directional pattern suggests the possibility that the prediction of various geologic aspects might well be improved, with the directional pattern serving as a common bond, linking together the diverse elements of a prospective area. Perhaps this would in turn improve the chances for success in oil exploration.

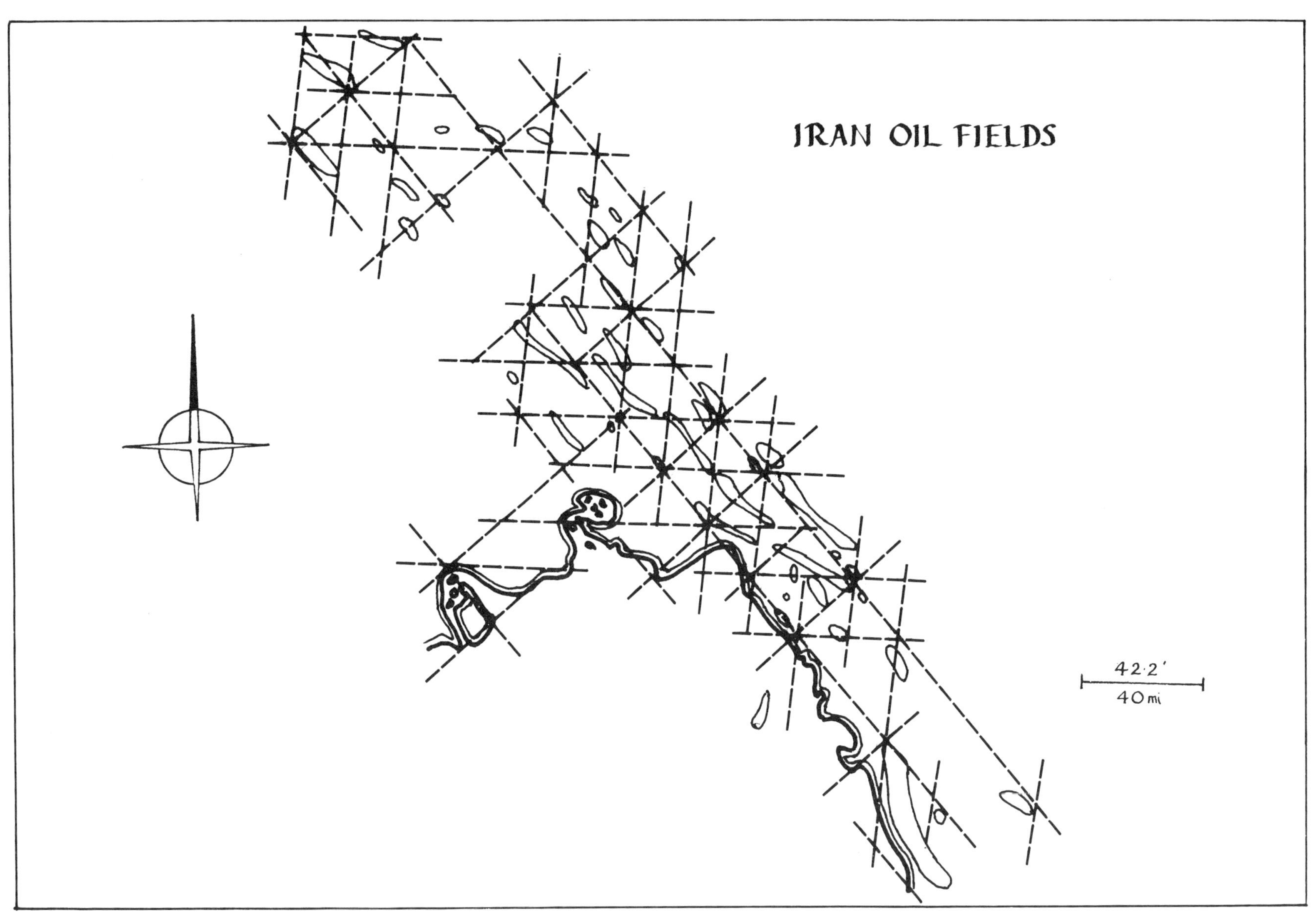

IRAN OIL FIELDS
42.2'
40 mi

6-4 *California Oil Fields*

This shows a group of California oil fields in Kern County, near the city of Bakersfield at the south end of the San Joaquin Valley. A number of fields in this area were discovered around the turn of the century and are some of the largest ever found in the United States. They include Kern River just right of center, and three others, Elk Hills, Buena Vista, and Mid-way-Sunset, that are located just west of those shown in the figure. However, large is relative, and the California fields are fairly small when compared with the Iranian fields that may contain five or ten times as much recoverable oil as the largest in California. The heavier lines in the figure show the outlines of the oil fields, and the lighter lines represent the surface traces of faults where the Earth has been fractured and torn apart.

A fault is a crack in the crust of the earth where the rocks on either side have been shifted in opposite directions. The shift may be a parallel displacement with the ground remaining level, or it can be vertical with one side of the crack moving up and the other side down. The stresses that act to tear apart the rock are many and varied, constituting a complicated and intricate system of forces. As with all natural processes, there has been an attempt to break up the system into a set of individual components that combine to produce the particular fault system observed. Although there have been varying degrees of success, no explanation has ever been developed that fully describes the relationship between the various stresses in the Earth's crust and the many diverse fault patterns created by those stresses. The problem is common to all natural processes, where not only is the information incomplete, but of necessity it is highly selective, and organized within a manmade set of categories. Each new item of information and each new idea may contribute to a better understanding of a natural process, but they also open more doors to the unknown. Where nature is concerned there will always be a wealth of unanswered questions, so that the solution of geologic problems will always call for trial and error.

The complex system of forces acting within the Earth's crust includes both the vertical gravity force that pulls rocks toward the center of the Earth, and the horizontal tension and compression forces that tend to pull rocks apart or jam them together. These forces work in combination, so that as the sides of a fault slide past each other, they are often also pushed against each other like two sheets of sandpaper rubbing together. This grinds up the rock on either side of the fault like grain is ground into flour between two mill wheels. Often, this produces a pasty clay within the crack that stops fluid from flowing across the fault, and this barrier can trap oil just as effectively as the structural folds in Iran. Of course, nature is not so easily categorized, and the oil reservoir traps are usually combinations of structural shape, sealing faults, and changing porosity. As always, nature refuses to conform to manmade categories.

Many of the oil fields in the figure are partially bordered by one or more faults that act as seals to hold the oil in the underground reservoirs. This makes for a very close correspondence between the oil field and fault system patterns. The general grain of the Kern County fault system is N40W, consistent with the overall trend of most other fault systems in California. Within this gross trend, a number of specific N40W directionals can be followed across the figure. The right half of the fault system is bordered on the northeast by a bundle of faults that lies between two parallel N40W directionals separated by a 5.27 min hierarchical interval. A N40W directional also traces the southwest sides of the three large oil fields in the center of the figure. In this same area are two east-west faults that lie on two W2N directionals, again separated by an equivalent 5.27 min interval. Below these faults in the lower right of the figure are three prominent faults, that trend N50E and form a right angle corner on the east where they intersect the N40W fault bundle. The right half of the fault system is divided from the left half by a N7E directional that strikes through the center of the figure, touching the ends of a number of faults. The left half of the system shows two prominent orientations. On the north two parallel faults follow two N40W directionals that are separated by a 5.27 min interval. On the south the grain is established by three W2N directionals, again separated by equivalent 5.27 min intervals.

It can also be observed that at center of the figure a number of faults seem to intersect in a node where several directional trends appear to converge. Moving clockwise around the node, the grain is N40W on the east, N50E on the southeast, W2N on the southwest and N40W on the northwest. There also seems to be a tendency for oil fields to cluster within each of these dominate trend areas. A closer examination might reveal similarities in these clusters that would help in the search for other productive areas. Of course, this discussion is highly generalized, and in a specific analysis much more detailed information would be studied in an effort to establish useful relationships.

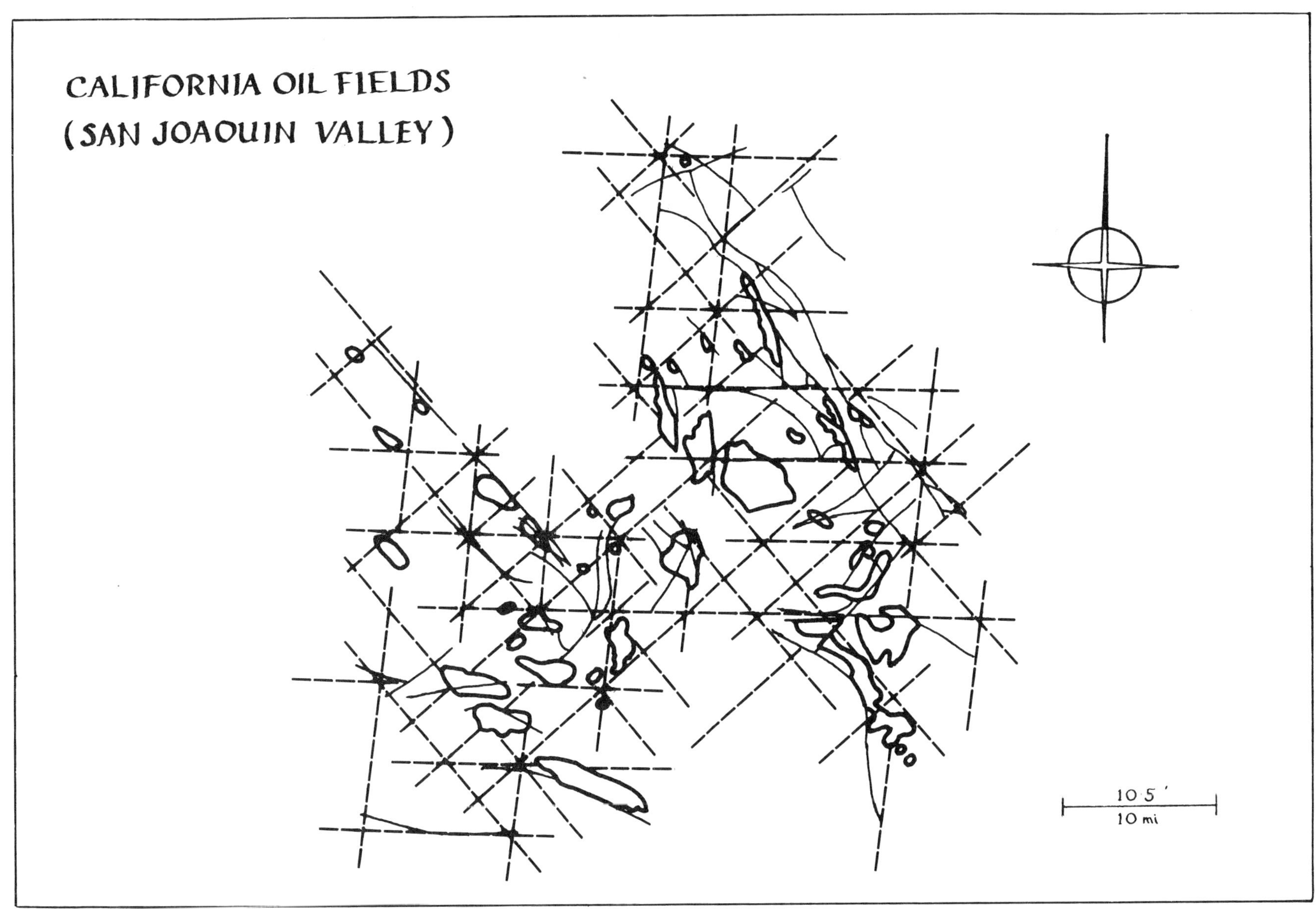

CALIFORNIA OIL FIELDS
(SAN JOAOUIN VALLEY)
10 5'
10 mi

6-5 *Texas Scurry-Codgell Oil Field*

This shows a geologic structure that contains the Scurry and Codgell oil fields in West Texas. The illustration was traced from Figure 13 on Page 197 in the the AAPG memoir 14 "Geology of Giant Petroleum Fields." The paper was written by E. L. Vest, Jr. and titled "Oil Fields of Pennsylvanian-Permian Horseshoe Atoll, West Texas." The Scurry Field on the south includes the Kelly-Snyder and Diamond M Fields, which are merely parts of a single reservoir that is one of the largest in the United States. The oil is trapped in this reservoir somewhat differently from those previously discussed. In Iran the oil was trapped under folds where the Earth's crust had been compressed and wrinkled up, and in California the oil collected behind sealing faults. In the Scurry-Codgell fields the oil is trapped in a string of porous limestone hills that were buried and sealed off by deposits of non-porous rock. This line of hills is called a "reef" and is made up of millions of small sea animal skeletons. The structure shown in the figure is only part of the full reef, which strikes off towards the northwest from both the north and south ends of the structure, forming a rectangle open to the northwest. When a reef is developing, it often parallels a coastline just offshore with its top at sea level. A good example is the Great Barrier Reef that lies off the northeast coast of Australia. Many tropical islands are circled by a reef that may be left as a ringlike atoll when the central island erodes away. Although the Scurry-Codgell reef may have been formed somewhat differently, it does have a semi-circular shape which gave it the name Horseshoe Atoll.

In the last two discussions the emphasis was on exploration, the search for new oil fields. Here, attention is centered on the analysis of an oil field that has already been discovered and is under development. An oil field lies deep underground, and information directly related to the reservoir comes from only a few wells about a foot across that have been drilled in an area that can cover many square miles. Other indirect pieces of information may also be available, but it all requires screening and extended interpolation before it can be used. From all these bits and pieces a possible picture of the oil reservoir is made to be used in estimating the economics of producing oil from the field. The picture would include the area and thickness of the oil filled rock, how much of the rock is pore space, how much of the pore space contains oil, how easily the oil flows through the rock, and how all these various characteristics are distributed within the reservoir. Rock can change drastically and abruptly over small distances, which is easily observed by walking along a rocky ridge. The pattern of these differences helps to determine the number of oil wells needed and where they might best be drilled. Cost estimates can then be developed which provide the basis for investment decisions.

The closed circular lines on the figure are contour lines, representing the depth below sea level of the Scurry-Codgell reef top. In the north end of the Scurry Reservoir the contours range from a depth of around 4,000 feet at the top of the reservoir down to 4,500 feet at the base, where the oil runs out and water fills the rock pores below that depth. The 4,500 foot contour traces the outer edge of the field, closing in the portion of the reef containing oil. The Scurry Reservoir is divided into two distinct sections that follow different directional orientations. The larger section on the southwest which trends N50E is relatively flat and broad, while the smaller section on the north trending N7E is a relatively steep sided reef mound. These two sections are separated by a N40W directional hinge that marks an abrupt change in trend. The southwest section is split in two parts by another N40W directional that lies 10.5 min west of the hinge directional. The north part of the southwest section is a wide bulbous area with a large appendage protruding out to the southeast, while the south part is more linear with smoother sides. Both parts span 10.5 min intervals between N40W directionals. Codgell Field at the top of the figure appears to follow a gradually changing trend from N7E in the south to N50E in the north. The overall impression of the entire figure reflects a N50E trend at the top and bottom, with a N7E offset in the center.

Just as with coastlines, oil field trends often show sudden shifts and offsets that can be important as an indicator of other relatively abrupt changes in the reservoir. This might be expected, considering the strain analogy in which the directional trends represent preferred flow paths. As previously discussed, deposits of various kinds from the flowing fluid may plug up the path, forcing the flow to abruptly shift to another directional trend. It follows that the place where the trend shifts may reflect some constriction that could reduce both the pore space that holds the oil and the ease of oil flow within the rock. The primary purpose of an expensive oil well is to find and produce oil. However, there is also a secondary purpose which is to obtain information that may help in future drilling. If a shift in the directional trend of an oil field indicates the possibility of a major change in reservoir rock characteristics, a well could be purposely located on the trend shift to test the possibility. Information from the well could prove very helpful in locating future producing wells. As in all specific cases, the usefulness of the directional pattern would depend on the degree of correspondence between the pattern and the available information. This in turn often depends on the intuition, interest, and ability of the analyst.

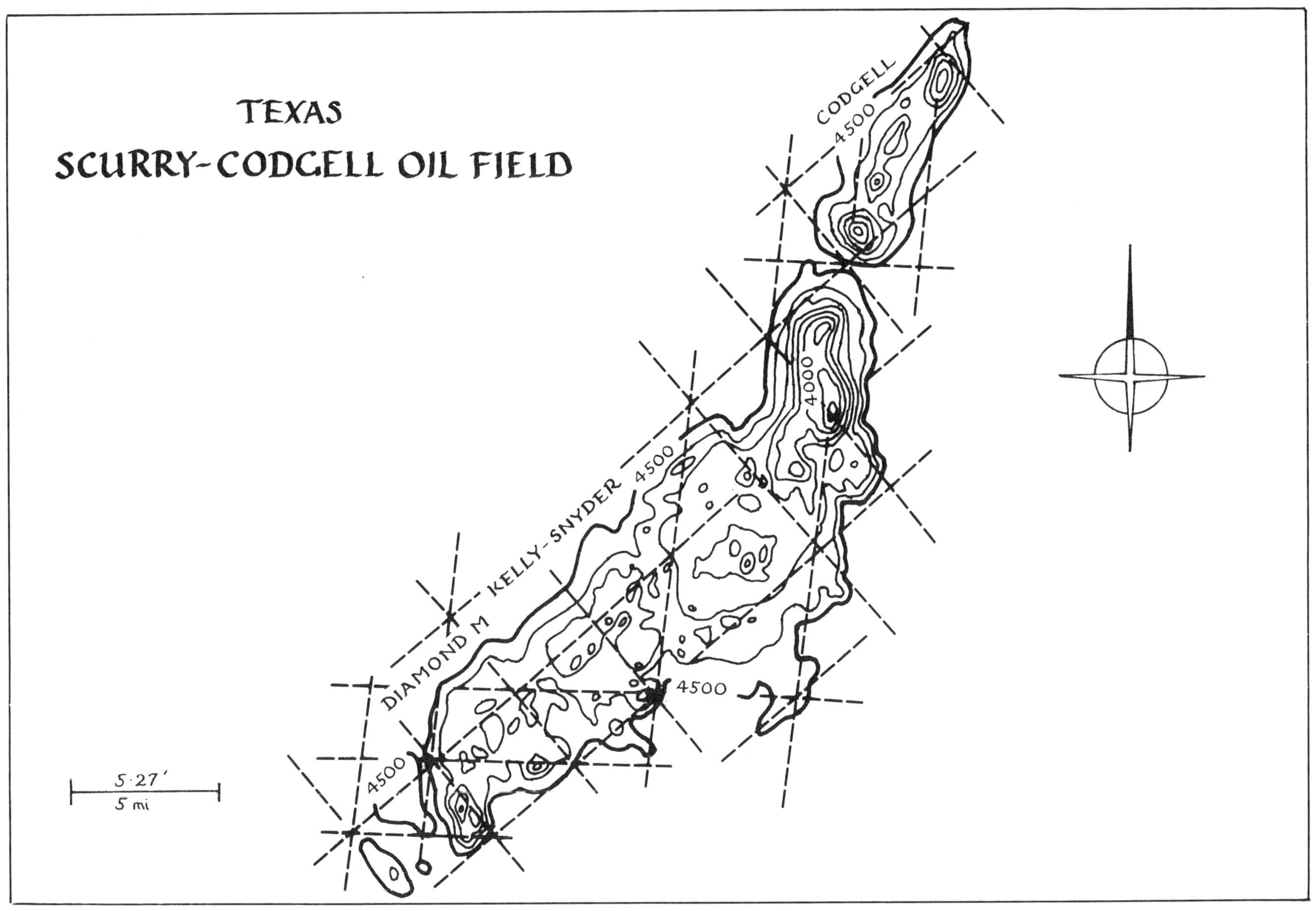
TEXAS
SCURRY-CODGELL OIL FIELD
CODGELL
4500
4000
KELLY-SNYDER
4500
DIAMOND M
4500
4500
5·27'
5 mi

6-6 *Netherlands Groningen Gas Field*

This shows the Groningen Gas Field in the northeast Netherlands. The illustration was traced from Figure 6, Page 365 in the AAPG "Geology of Giant Petroleum Fields," from a paper by A. J. Stauble and G. Milius. The field is one of the largest in the world and conveniently located close to several highly industrialized areas of Europe. It lies just below the North Sea, and it was the Groningen gas discoveries that supplied the incentive to drill for petroleum further north in that area. The gas in Groningen is trapped in porous sandstone below a bulge in the Earth's crust that is sealed above by layers of salt and other non-porous rock. The sandstone reservoir is split up by a complex pattern of faults that gradually disappear above the reservoir where they run into the plastic-like salt seal.

When petroleum is produced it flows out of the underground reservoir pores and into a well that carries it up to the surface. This happens because the fluid tends to move from the high pressure underground to the low pressure on the surface. But there can be no empty space underground, so when oil or gas leaves the rock pores something must come in to replace it and keep the pores full. Often the replacement fluid comes from the oil or gas left in the reservoir that expands to fill the pore spaces as the reservoir pressure drops. Sometimes water from rocks around the reservoir flows in to replace the produced fluids. When nature is uncooperative in refilling the pore spaces, water or other fluids may be pumped down from the surface through injection wells and into the reservoir. The flow patterns involved in this process of producing and refilling reservoir rock over a large field area are highly intricate and complicated. The pattern depends on both the distribution of rock flow characteristics and the location of both the producing and injection wells. The pattern of rock characteristics forms a network of preferential flow paths through the pores that can be complicated by a fault system in which individual faults may sometimes help and sometimes hinder the flow. Fluids move through this intricate maze of preferential flow paths interlaced with flow constrictions as they travel towards a producing well, or away from an injection well. As new wells begin to produce and production is cut back in other wells, the distribution of production continually changes, adding another complication. It seems clear that the analysis of changing flow patterns within a reservoir presents a formidable problem.

With the advent of computers, much of reservoir analysis is now based on simulation models, which are merely mathematical analogs of nature used to calculate producing behavior predictions. It is sometimes forgotten that the mathematical models are only very rough representations of the actual intricate and complex reservoir. As with all geologic analysis, it is necessary to isolate and subdivide the segment of nature under study into some organized system. In a simulation model the physical reservoir is usually represented by a rectangular array of squares or blocks that cover the field area. Each of these blocks is considered a discrete piece of the natural reservoir, with no variation in the rock and fluid it contains. Of course, this is only a very gross picture of the highly variable natural reservoir. A system of fluid and flow equations is then set up that interconnects the blocks in the array. Production of petroleum is represented by reducing the amount of oil or gas in a particular block where one or more producing wells are located. This sets off a mathematical trial and error calculation that adjusts the pressure and amount of oil, gas, or water in each block until the results fit the system of equations. The initial results of the calculations rarely match the actual information collected from the producing wells, so that a second trial and error process is required to bring the calculated results in line with actualities. In this second calculation the assumed rock and fluid characteristics are adjusted, and the calculations repeated until a reasonable fit is obtained.

A major problem in this double trial and error solution is the presence of up to thousands of blocks, each with its own set of reservoir and fluid characteristics, any one of which could be adjusted. Therefore, a good match between actual and calculated values can usually be obtained in numerous ways, and rarely is any one way unique. This general problem is common to most geologic analyses. Given an actual observation, there are almost always a variety of logical event chains that will describe a reasonable geologic evolution leading to the observation. In simulation studies it is common practice to run predictions with several sets of adjustments and compare the different results to show the uncertainty involved. However, it is usually necessary to select a best guess prediction that will be used for economic forecasts. The validity of that forecast is highly dependent on the selection process used by the analyst to subdivide the reservoir into blocks, and to adjust the various characteristics to obtain a match with actuality.

The directional pattern might prove useful in this process by providing a more logical framework for dividing up the reservoir into blocks. The pattern seems to be linked to natural processes, and its hierarchical nature allows for a coarse or fine gridwork pattern. The diagonal directionals representing peak strain lines are conveniently at right angles, and fault systems appear to fit well within the pattern, as shown here in Groningen Field. In addition, with the block subdivision connected to a natural Earth pattern, both initial estimates and final adjustments of the block characteristics might well be improved to the benefit of the forecast.

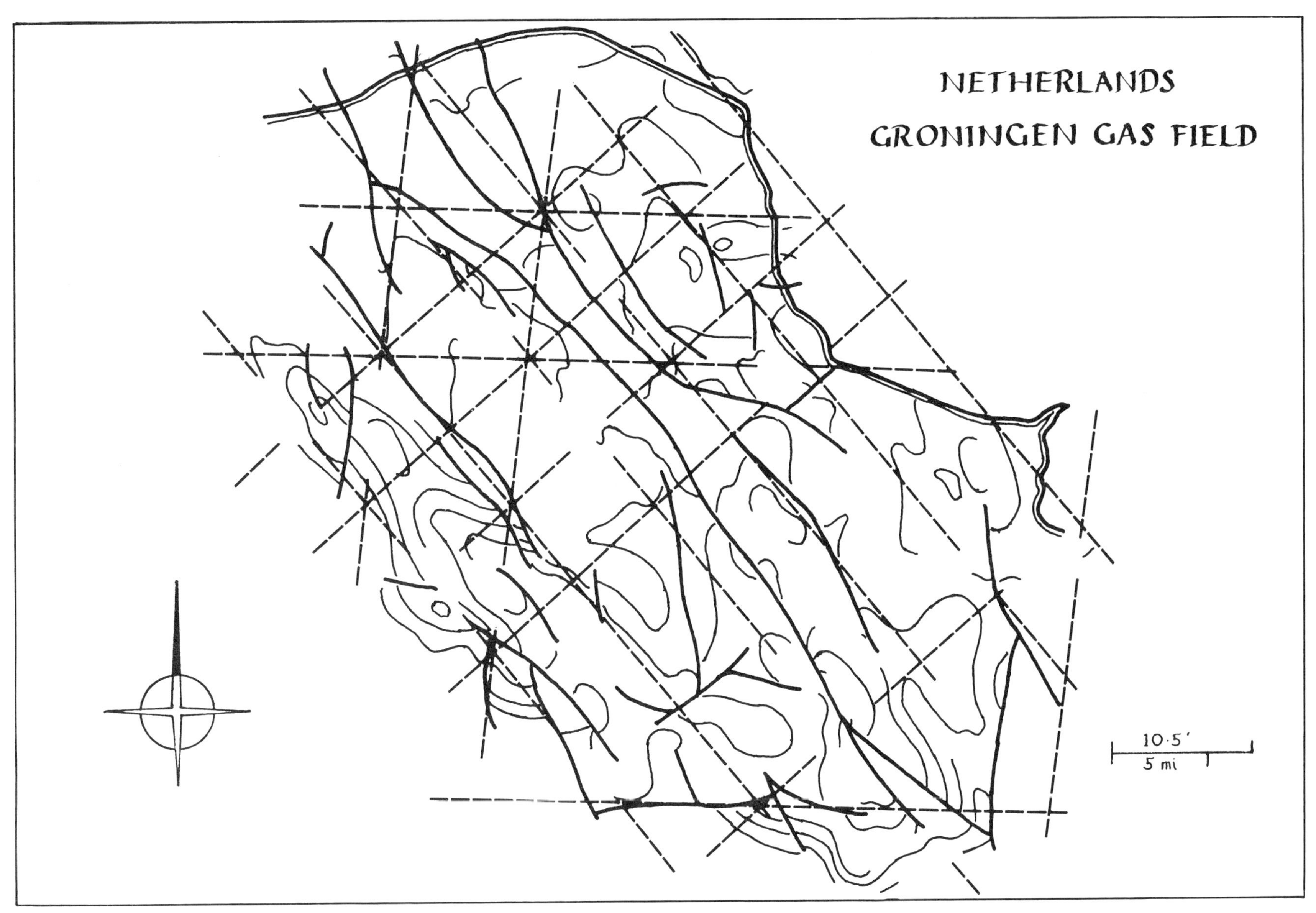
NETHERLANDS
GRONINGEN GAS FIELD
10·5'
5 mi

6-7 England Northern Pennine Ore Field

This shows veins of ore in the northern Pennine mining fields around the English town of Alston, located just south of the Scottish border on the spine of the Pennine Mountains, that run north and south and rise up to almost 3,000 feet. Alston is near the center of the figure, and it is cut off from the surrounding area by a high moorland plateau. Lead was mined in this area from Roman times up to the end of the 19th century, when cheaper imports closed down the mines. Alston claims to be the highest market town in England, and just a few miles away is Nenthead at an elevation of 2,000 feet, which was laid out in the last century as a model village for the lead and zinc miners in the area.

Lead has been an important metal since prehistoric times. The Egyptians were using lead 7,000 years ago, and the Phoenicians mined lead in Spain 5,000 years ago. The ancient Romans used lead pipes to carry water, and set up what may have been one of the first sets of manufacturing specifications. The lead pipes were made in standard diameter sizes and all were ten feet long. The Latin word for lead is "plumbum," the source of our words for plumbing and plumber. Before Rome occupied Britain in the middle of the first century, lead and other metals were exported from the mineral rich British Isles to the Roman Empire. It seems probable that a principal reason for the Roman conquest of England was to gain control of the mineral sources. This also may have been the prime reason for the construction of the wall by the Emperor Hadrian just north of the Pennine ore fields, which protected the mines from the Scottish raiding parties coming down from the north.

The faults that carry the Pennine ore veins cut across a limestone formation that covers much of the English uplands, and is appropriately called the "Mountain Limestone." This thick limestone formation was formed more than 300 million years ago during a time called the "Carboniferous" when shallow seas covered large areas of the Earth. Over millions of years the waste from small sea animals piled up on the sea floors, forming the thick limestone layers that later rose above sea level and are found today in many parts of the world. It forms the steep cliffs of the Redwall Limestone in the Arizona Grand Canyon and the rectangular blocks of rocks that top the Canadian Rockies above Banff in Alberta. Mount Everest in Asia, the world's highest peak, is also capped by this widespread limestone formation. It is thought that during that same Carboniferous time, hot salty water carrying minerals oozed up from deep in the Earth's crust and forced its way to the surface. As the saturated solutions approached the surface, both temperature and pressure dropped and mineral crystals solidified out of the solution. The same type of process can be seen today in areas such as Yellowstone Park, where hot volcanic rocks below the surface heat water that is saturated with mineral salts. The hot brine bubbles up at the surface in hot springs or geysers where the colored minerals come out of solution and coat the nearby rocks. It is these mottled red and yellow layers of mineral deposits that gave Yellowstone its name.

After the Mountain limestone had been laid down and the scattered patches of rich concentrated mineral deposits formed, the limestone was squeezed and wrinkled up to form the Pennine Mountains. The enormous forces that acted on the rock not only wrinkled it, but also cracked it in many places to form the extensive fault system seen today. While all this was taking place, the minerals were in some way driven out of the scattered patches and redeposited in the fractures of the fault system to form the rich veins of ore. As with all geologic explanation, there is much in the way of supposition due to the incomplete information available. As Earth process occur, they tend to wipe out the effects of earlier processes, so that the further back in time, the less information is available. There are two distinct phases in the development and analysis of ore veins. First comes the exploration, or prospecting, where the search is for new ore bodies, then comes the need to estimate the size and continuity of an ore vein once it is found. It is not unusual to discover a rich ore vein and then have it fade away after money has been spent on the mining operation. In the type of ore deposit shown in the figure, an analysis of the ore bodies first requires an analysis of the fault system.

As in other fault systems, the directional pattern provides a framework for the fault pattern. Many of the shorter faults end on directionals, while others break midway between directionals, maintaining the hierarchical relationship of the directional pattern. In the upper left part of the fault system is a group of larger faults that form a distinctive rectangular pattern orientated N50E and N40W. Several of the faults in this group lie on or midway between directionals. Just below this fault group the grain appears to follow W2N within a band that cuts across the system, dividing it into north and south parts. The directional pattern could provide a base for organizing the fault system, which might allow for better predictions of the extension of known ore veins and the location of new veins in the immediate neighborhood.

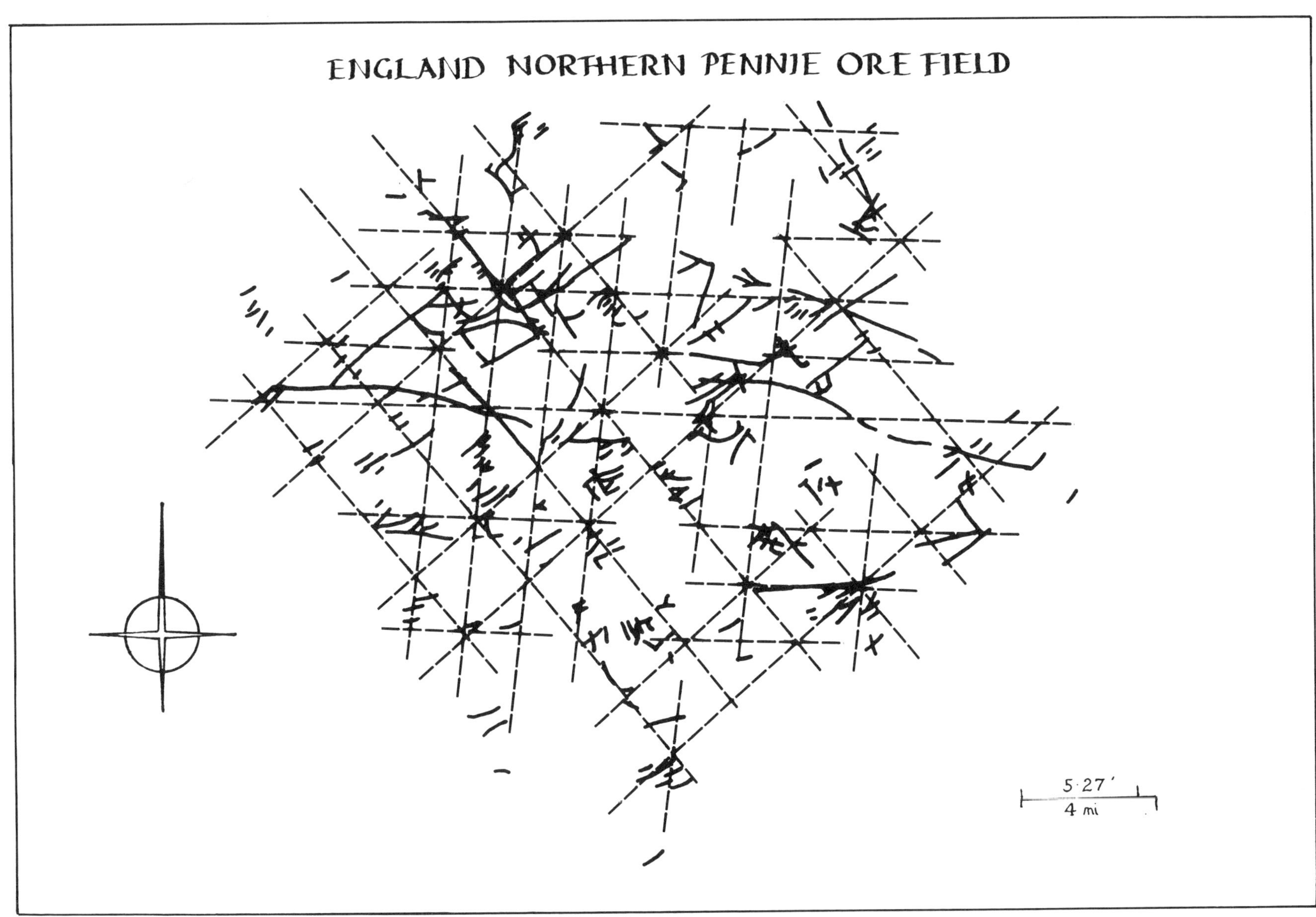

ENGLAND NORTHERN PENNIE ORE FIELD
5·27'
4 mi

6-8 South Africa Witwatersrand Gold Field

This shows the surface traces of the gold reefs in the South African Witwatersrand Basin that lies at the center of one of the greatest concentration of mineral deposits in the world. The basin is only about 20 miles across and 100 miles long, but it has produced more gold and has still more gold to produce than any other area in the world of comparable size. It is estimated that it contains half of all producible gold in the world. A few miles to the northeast is the Bushvelt Complex that contains over half the world's reserves of platinum and chromium, and has the largest production in the world of these two metals. To the southwest lie the Kimberlite pipes that have the largest production of gem diamonds in the world. It is thought that the Witwatersrand gold probably came to the surface and filled fractures in volcanic rocks about three billion years ago during the Precambrian Age. Possibly a half billion years later the rocks were torn apart, and gold from the fractures collected in beds of broken rocks called "conglomerates" that extend down to great depths. Some of the gold mines in the area produce ore from shafts more than two miles below the surface.

It was the discovery of diamonds and gold during the late 1800's that started the rapid development of South Africa. In the mid 1600's the Netherlands set up a ship supply station on the Cape of Good Hope at the southern tip of Africa. Dutch farmers called "Boers" were brought in to work the land and produce the food that supplied the ships. Over the next 150 years the Boers gradually moved inland as the population grew, in much the same way British colonists on the North American eastern seaboard moved west toward the Appalachians during the same period. In 1806 after Napoleon was defeated at Waterloo, the Capetown settlement was handed to Britain, and British colonists began to come into the area. Naturally, there was conflict between the new British settlers and the wandering Dutch farmers

called "Trekboeren" who had lived in the area for a century and a half, so the Trekboeren moved north in the Great Trek to settle in the Transvaal, north of the Vaal River in the high velt. They found a devastated, empty land of destroyed villages and starved refugees running from a conflagration approaching from the east. This was the Zulu nation under King Shaka that had conquered Natal on the east coast, and was moving west into the Transvaal just as the Trekboeren were moving north into the same area.

Shaka was born an illegitimate son of a Zulu king, and at an early age was exiled with his mother to a different tribe to grow up saturated with hatred and revenge. When his father died he manipulated his way to become king, and then proceeded to have his revenge. He created an army designed to operate as his own private force of destruction and began the extermination of all who were even remotely connected with his exile. When new territory was conquered whole villages were destroyed with all people tortured and killed, so that by the early 1820's all Natal was in ruins and refugees were streaming west into the interior, completely disrupting the tribal structure in all of South Africa. It is estimated that more than two million natives were killed in the widespread conflicts. Shaka ruled by fear and terror, and each day would pick at random a few of his own people to be killed. In the late 1820's his mother died, and Shaka called for one year of mourning which began with the indiscriminate killing of thousands. No crops could be grown, no milk drunk, and all women who became pregnant were to be killed along with their husbands. This completed the land's destruction as the search for food spread out further and further. Although Shaka was eventually killed by his own people, the damage was done and the Trekboeren moved in from the south to claim an empty land, virtually devoid of native people.

In the 1860's the rich Kimberlite diamond pipes were discovered, and this gave the British the incentive to force their way into the region and eventually to annex the Transvaal. Then in the 1880's the Wit-

watersrand gold reefs were discovered and the Rand gold rush began. This was one of several 19th century gold rushes beginning with a quarter million 49ers pouring into California over a five year period. A few years later gold was discovered in Australia, and a half million from the British Isles swarmed into the country. Then came New Zealand in the 1860's, South Africa in the 1880's, and the Yukon at the turn of the century. In all cases disruptions were severe, which in South Africa led to the Boer War that ended with the British in complete control. Gold has certainly been a major factor in cultural change throughout the world over the last century.

The pattern of the Witwatersrand gold reefs shown in the figure may have come from a group of deep cut valleys that were later filled with conglomerates rich in gold. The general trend is N40W, with a shift to W2N towards the north. Several N7E and N50E directionals that lie along the ends of reefs may be traced across the figure. Based on previous examples this is characteristic of major fault systems, so that fracturing may have been involved in the formation of the reefs. Whatever the process, the reef pattern fits well within the directional pattern, providing an organizing device as in other examples. This could help in predicting extensions in the individual reefs.

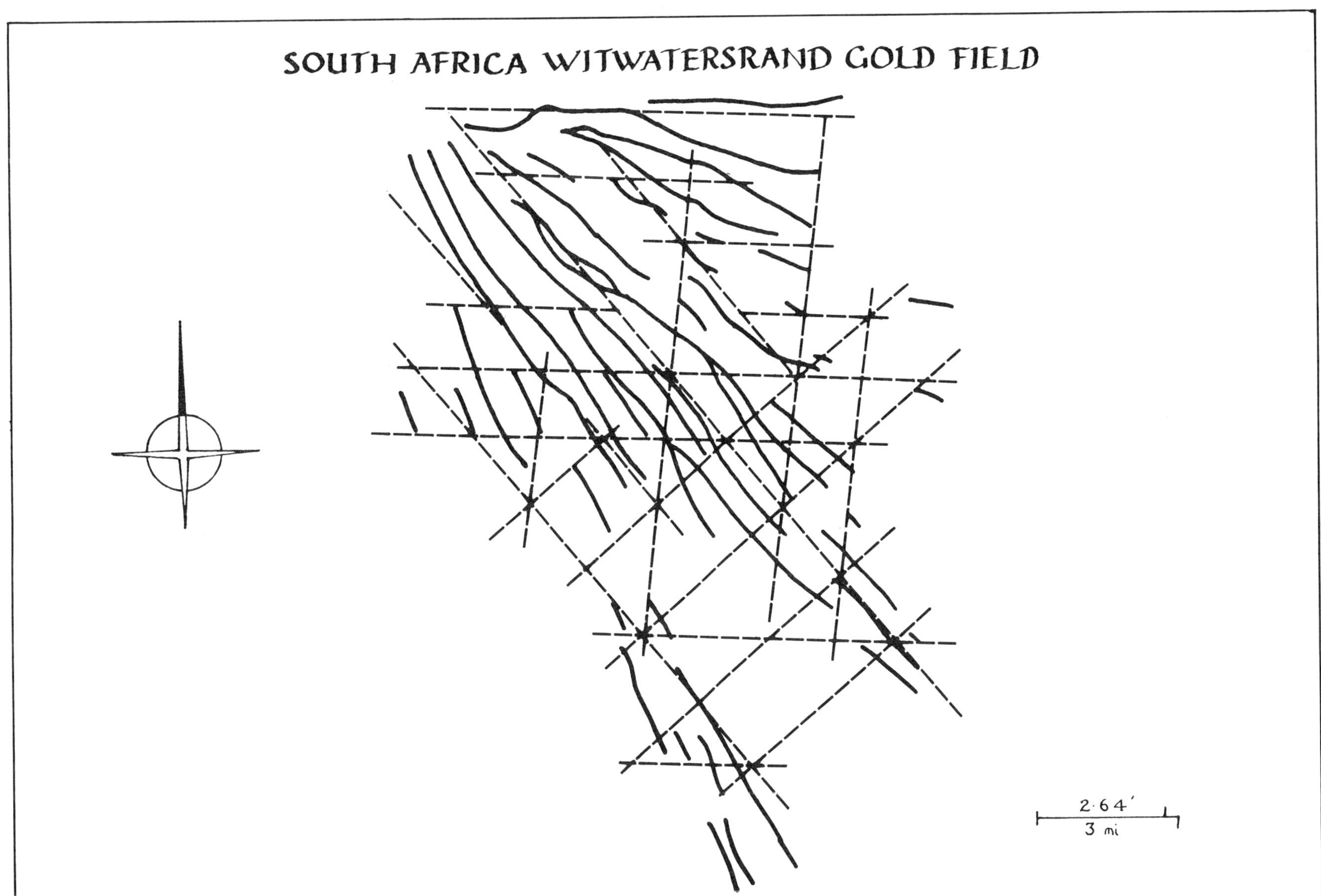
SOUTH AFRICA WITWATERSRAND GOLD FIELD
2.64'
3 mi

6-9 California Geysers Geothermal Field

This shows the surface fault pattern above the Geysers Geothermal Field. The illustration was traced from Figure 7.2, Page 207 in the book "Geothermal Systems" by L. Rybach and L. J. P. Muffler, from Chapter 7 written by Max D. Crittenden, Jr. The field is located just north of San Francisco, California in Sonoma County where some of the best California wines are produced. The Geysers is the largest geothermal field in the world and produces steam from wells drilled as much as a mile below the surface. The area has always been known for hot water features such as geysers and hot springs, and in the 1920's a few shallow wells were drilled in an attempt to find large steam flows. However, it was not until 1950 that deeper wells found steam in commercial quantities, which led to the first small power plant in 1960 that produced electricity from the steam. Since then, there has been continuous development with enough electricity generated today to supply all of San Francisco.

A geothermal operation is somewhat different from mining ore or producing petroleum. In mining and petroleum production a physical raw material such as solid ore, liquid oil, or natural gas is taken out of the ground and processed to obtain a commercial physical material such as metal, gasoline, or plastic. The commercial material remains a physical product, and in essence is merely a part of the physical raw material that was originally produced from the mine or well. In a geothermal operation the wells produce hot salty water or steam that condenses to water. The produced water has no value as a physical material, but only serves to transport heat energy from below ground to the surface. The heat energy is released when the steam condenses or hot water cools, and is then used to heat buildings or generate electricity. All of the produced physical material ends up as cool, salty waste water that not only has no value, but costs money to get rid of. In effect, a geothermal operation mines heat energy from underground.

Temperature increases almost everywhere as depth increases, but there are localized places where it is much hotter than usual at a particular depth. Generally, this happens above a large underground mass of very hot rock that forced its way up from deep in the Earth as liquid rock, or magma, and became solid before it reached the surface. As the magma rose towards the surface, it broke up the rock ahead of its path as it forced its way through the Earth's crust. This created a fracture system above the hot mass of volcanic rock. Both hot water and steam are lighter than cold water, so that a circulating system would tend to form above the hot rock with cool water seeping down and hot water and steam bubbling back up through the fracture system. The hot water and steam that rises to the surface comes out in hot springs, bubbling mud ponds, or fountaining geysers. In the usual geothermal operation, wells are drilled deep into the crust above the hot mass of rock to produce hot water or steam from the bottom of the circulating system. With the production of hot fluid from the wells, the hot springs and geysers often dry up and eventually die out completely, because more hot fluid is being taken out of the system than nature can replace by slow seepage from the surface.

The Geysers Geothermal Field is somewhat different from the above description. There appears to be a partially molten mass below the field that has heated the water seeping down from the surface and turned it into steam, which tends to bubble up towards the the surface. However, much of the rising steam has been trapped in a reservoir below a layer of nonporous rock, just as petroleum is trapped below a rock seal. Evidently, the steam has been collecting in the reservoir for many years, making it the largest geothermal field in the world. The heat energy in the reservoir is not only contained in the steam, but also in the rock, and some of this rock heat is being produced by injecting cold water into the reservoir which heats up when it contacts the hot rocks and is then produced to recover the heat energy.

The lines on the figure show traces of fractures that have cut the surface, and these were mapped to determine landslide-prone areas after a number of wells were destroyed in small local landslides. The fracture system corresponds to the directional pattern much like other systems previously discussed. Consistent with most of California, the general grain trends N40W along two major fault zones, Collayomi and Maacama, that border the system on the northeast and southwest. These fault zones coincide with two N40W directionals, and between them are two more N40W directionals that divide the system into three 5.27 min strips. The field itself is contained within the upper two of these strips across a 10.5 min interval. Several other directionals can be traced through the system that mark the ends, breaks, and direction changes of individual faults. Based on the shear analogy, the directionals seen on the surface might also reflect a deeper fracture pattern in the reservoir, because the entire crust of the earth would be involved in the torsional stress. This does not mean that the fracture pattern would be unchanged at various depths, but rather that the same directional framework could be used to organize the different fracture patterns. In certain cases this might allow surface features to be used in developing a directional pattern that would reflect the pattern of deeper seated geologic features.

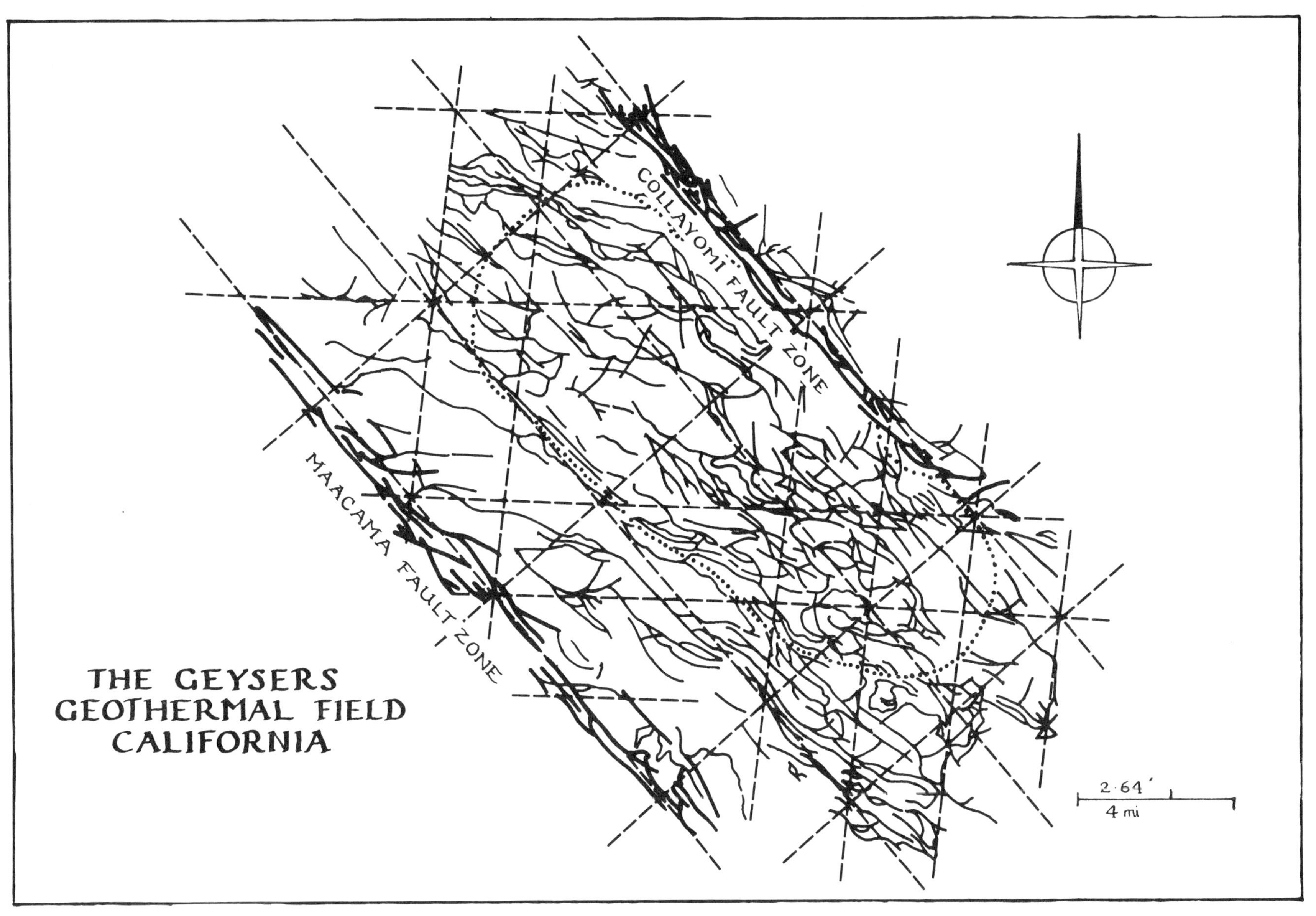
COLLAYOMI FAULT ZONE
MAACAMA FAULT ZONE
THE GEYSERS
GEOTHERMAL FIELD
CALIFORNIA
2·64'
4 mi

Several examples have been presented to illustrate how the directional pattern might be applied in practical geologic analysis. Examples were shown of stream channel shifts, petroleum fields, ore deposits, and a geothermal field. In all cases the directional pattern provided a framework for the organization of various geologic features. In geologic analysis the available information is often scanty and scattered, and it requires knowledge, experience, and imagination to fill in the gaps with best guesses. The directional pattern could provide a guide for the many estimates that must be made to complete an analysis. The pattern is very flexible due to its hierarchical nature, which allows the network intervals to be rationally and systematically adjusted to accommodate a broad range in scale, coverage, complexity, and completeness. This could not only improve predictions in many cases, but would also provide a more rational and understandable approach to the planning, review, and presentation of an analysis.

There are many questions to be answered in formulating a logical and reasonable approach to an analysis. One question, that sometimes receives considerably more emphasis than warranted, concerns the importance of the subject matter. This concern often plays a dominate role in the practical decisions that must be made when allocating grants to the scientific community. However, there may be little rational in funding research on some very important matter if it is doubtful that the necessary basic information can be developed. Planning requires that available information be critically studied to determine the practical possibilities of success before the major effort begins. An understanding of the quantity and quality of available information provides the basis for comparisons of alternate approaches to the proposed analysis. From this come estimates of the time and money required and the limitations expected in the end results. The last item is of major importance and is often neglected in the initial planning. If an analysis comes up only with non-specific generalizations that provide little guidance, the value may not be worth the effort. In the practical world, specific problems require specific direction, and worthwhile analysis planning must take this into account.

The directional pattern could be helpful throughout the planning process. Information can be set into the framework to form a picture of the analysis base, and where information is lacking, gaps would appear in the pattern. Anomalous bits of information might also be more readily visualized as irregularities, and this would bring attention to certain items that may require more concentrated study. Increased understanding almost always comes from the study of information that does not fit the norm. All too often, data that does not conform to a preconceived pattern is thrown out and ignored in an analysis, based on reference to a statistical comparison test that uses some arbitrary standard. Selectivity of information is vital in all geologic analysis, but it requires a rational geologic basis. It is important that those items that do not fit and seem unexplainable are singled out to give others, with a different viewpoint, a chance to reexamine the problem. With the geologic information in place, the directional pattern would serve as a nucleus for discussion of alternate approaches. There is no standard, routine procedure in analysis. Multiple sets of data may be examined in a multitude of ways, that can vary from a simple verbal discussion to a complicated and complex mathematical model. The methods used depend on many things, including the quantity and quality of available information, the purpose and objectives of the analysis, and the availability of technicians, equipment, and time. A visual and easily understood organizational system with emphasis on possible interconnections within the pattern could help a great deal in these preliminary discussions.

Much of the process of planning and review is presentation, which is essentially a form of communication between the analyst who does the study and the end user who must eventually make a decision based on the study. The basic approach of these two is somewhat contradictory. The analyst is necessarily concerned with a multitude of detail and the resolution of many inconsistencies. His work is never really complete, but must stop at some point based on an intuitive feel that any remaining inconsistencies are relatively unimportant. Of course, intuition is rarely definitive, so the analyst tends to review the unresolved questions repeatedly in an attempt to tie up one more loose end. However, the end user is usually not interested, and rarely has the time to concern himself with details. He would like to quickly find a specific answer to a specific problem and then move on to another problem. He would prefer not to be bothered with the analyst's frustrations, and tends to override or pass off the many hanging inconsistencies the analyst wants to highlight. These differences in outlook often cloud presentations of progress reviews and study results. The consistent use of a common organization scheme in all presentations could help bring together the diverse attitudes of those concerned with a study.

This concludes the broad suggestions of possible pattern use. Any specific application would require much trial and error to develop an understanding of the constraints and limitations of the pattern correspondence. This means that first attempts at using the pattern will probably require considerable time.

7 THE CRITICISM

QUESTIONING THE PATTERN

Ever since the world was mapped, people have commented on the regularities of continental coastlines. The matching shapes of the coastlines on either side of the Atlantic were noted by the scientist and world traveler Alexander Humboldt at the beginning of the 19th century. In the 1850's the Italian-American Antonio Snider-Pelegrini published a map with no Atlantic and the continents on either side fitted together. Others saw geometric patterns around the globe. Elie de Beaumont in the first half of the 19th century saw sets of regular polygons on the Earth's surface, and he eventually decided on a network of 20 sided icosahedrons within 15 great circles. In the last half of that century, William Louthian Green, who served twice as Prime Minister of the Hawaiian Islands, wrote three volumes on the tetrahedral earth. More recently in the mid 20th century, Samuel Warren Carey visualized a hierarchical pattern of polygons that covered the earth. For the most part, these studies were subjective observations combined with a very imaginative but somewhat speculative discussion, and as such were open to criticism from the scientific world. The observations and explanations discussed here are also open to criticism because of the many inconsistencies that remain unresolved. In this chapter a few of the major discrepancies associated with the directional pattern and strain analogy are discussed. As yet there are no answers to the questions raised, so that the discussions do little more than point to possible areas of future study.

It has been emphasized throughout that the material presented is essentially an empirical observation in which a number of Earth features on Mercator Projection seem to fit into a pattern made of four orientations called directionals. These four directionals were observed to be spaced at regular intervals within a doubling series, forming a hierarchical framework made of the sides and diagonals of rhombus figures. The rhombus' uprights were inclined north 7 deg east and the laterals west 2 deg north, with the diagonals that stretched between opposite corners following north 40 deg west and north 50 deg east. These four directionals were called N7E, W2N, N40W, and N50E, equivalent to the observed orientations. This observed pattern on Mercator Projection was found to resemble the pattern of extreme strain lines created by twisting a thin walled cylinder and then unrolling it. The twisted cylinder fitted around a twisted Earth. The observed hierarchical series of intervals could be matched by repeatedly halving the 360 deg circumference of the Earth. The directional orientations could be closely approximated from elements of the one-half series by taking the inverse tangent of one-eighth to give 7.2 deg and one thirty-second to give 1.8 deg. The diagonals of a rhombus constructed with these angles are computed at 40.5 deg and 49.5 deg. The correspondence between the strain rhombus and directional pattern is also close in relation to the observed degree of clarity between directionals. The strain rhombus diagonals that represent peak strain correspond to the two directionals that show up most clearly on the Earth, and the top and bottom sides of the rhombus that represent least strain correspond to the W2N directional that shows up least clearly.

An analogy may correspond well with certain selected observations, but that, of itself, is certainly not proof. One way to think of an analogy is to consider it a simple model that broadly corresponds to nature in some major aspects, but is deficient when examined more closely. It is well known that people tend to see regular patterns in irregular groups of items, and as noted above, many patterns have been suggested that seem to reflect the patterns of various Earth features. Statistical tests rarely give much help in determining the best pattern because they require comparison with a somewhat arbitrary standard. This problem of alternatives will be illustrated by presenting a different pattern that seems to fit about as well as the directional pattern discussed. Another problem that is common for many aspects of nature is the boundary problem that concerns the constraints that define the limited real world in which the analogy will work. These limits, or boundaries, describe the type and size of objects considered as well as other physical constraints. The possible lower limit of the hierarchical rhombus dimensions will be discussed, and one additional example of a somewhat different Earth feature will be presented.

Two other concerns are presented. Plate Tectonics is the newly accepted, all pervading theory today, and any and all new geologic observations or ideas are expected to conform with that theory. Plate Tectonics is a dynamic theory that deals with Earth features that shift and change with time, while the directional pattern represents a static description of today's world. Although a direct comparison between a dynamic theory and a static description is at best difficult, the possible relationships will be discussed in a general sense. The second concern deals more directly with the strain analogy, in which the Earth is twisted through an exact 7.2 deg angle. It is difficult to imagine just how a torsional stress might act to accomplish this. In the last example that ends the chapter and the book, an enigma is presented that shows a close correspondence between the directional pattern and an ancient man made megolithic structure.

7-1 Zagros Fault System II
Is There An Alternative?

This shows the Zagros Fault System of Iran that lies just north of the Persian Gulf and overlies the Zagros Mountain Range. In chapter 4 it was shown that the directional pattern provided a framework for the fault system, with the ends and breaks of the individual faults lying on the pattern directionals. However, the orientation of the relatively long and straight Zagros Main Fault does not follow any one of the four directionals. The measured trend of this fault is north 50 deg west, representing a 10 deg counterclockwise rotation from the N40W directional. The Main Fault orientation can be called N50W, and it just happens to be symmetrical with the N50E directional, so that it might well be included in an alternate pattern that is a mirror image of the original directional pattern. The alternate pattern is formed by shifting the original rhombus uprights from N7E to N7W, and the laterals from W2N to W2S, which shifts the rhombus diagonals from N40W to N50W, and from N50E to N40E. In this figure the original directional pattern has been replaced by the mirror image alternate with the same 2.81 deg interval, and as with the original pattern, it shows a close correspondence to the fault system.

This alternate pattern has been tested in other areas, and often fits the natural features almost as well as the original directional pattern. A number of continental coast lines seem to conform even more closely with the mirror alternate than with the original. The southeast coast of South America trends N40E, and the southwest coast of Africa tends to follow N7W. If offsets are ignored, and a mean straight line is drawn along the eastern seaboard of the United States, the line trends N40E. The northern half of New Zealand's South Island, shown in chapter 1, also trends N40E, and in chapter 3 the southern Mauna Loa lava flows seem to fit the alternate N7W more closely than the original N7E. Other examples can be found on other planetary bodies. The Mars Tharsis bulge that was discussed in chapter 5 is a broad area considerably higher than the surrounding surface with four of the largest of Mars' volcanoes on its crest. Three of these volcanoes, Arisa Mons, Pavonis Mons, and Ascraens Mons, are equally spaced along a N40E line that forms the crossbar of a tilted "T." The "T" upright is a N50W joining the middle volcano Pavonis Mons to Olympus Mons in the northwest, with the upright length about equal to the crossbar length. Many other features on Mars fit well within the mirror alternate pattern.

The existence of two symmetrical patterns that both fit selected portions of nature is not necessarily in conflict. Nature is infinitely complex and may be described in a multitude of different ways. It is not unusual to find that two seemingly divergent ideas eventually merge into one somewhat different pattern. An extended understanding of nature often arrives in the form of reconciliation of what appears to be unreconcilable observations. An explanation or theory is designed to be compatible with selected categories that provide the boundaries to validity and usefulness. This may well be a major problem in statistical testing where mathematics attempts to match a proposed pattern with a selected set of observations. It is not unusual that the resulting disagreements are directly related to the selection criteria differences. However, with regard to the original and alternate patterns, there is at least one way to explain the difference and still retain the stress analogy.

In the stress analogy the rhombus strain pattern formed by twisting a thin-walled cylinder was found to resemble the observed directional pattern on the Earth. Although it was implicitly assumed that the twist was counterclockwise when looking down on the north pole, it might just as well have been clockwise, which would give the alternate pattern. This leads to the possibility that the twisting stress is cyclic or oscillating, so that the earth is periodically twisted first one way and then the other. This could be compared with a clock pendulum that swings back and forth. Imagine the pendulum started by holding it up on the left side and letting it go. As it drops, it picks up speed, going faster and faster until it reaches the vertical where it then begins to slow down as it moves up on the right side. Eventually it slows to a stop, hesitates, and then changes direction to swing back down toward the vertical, again picking up speed. As applied to the Earth, the process would start from a maximum clockwise twist of N7W. As the Earth unwound like a spring, speed would increase until the counterclockwise twist was gone and the clockwise twist began, then the speed would decrease until the twisting stopped at N7E. At that point the process would be repeated as the earth began to unwind in the opposite direction.

If something comparable to the above is occurring, it might explain the directional pattern and its mirror image. They could represent an impression of the strain pattern at the two twisting extremes when the motion comes to a standstill before reversing direction. At all other times the strain pattern would be changing, and consequently blurred. Of course, there is at present no indication of this, so that the argument is little more than speculation.

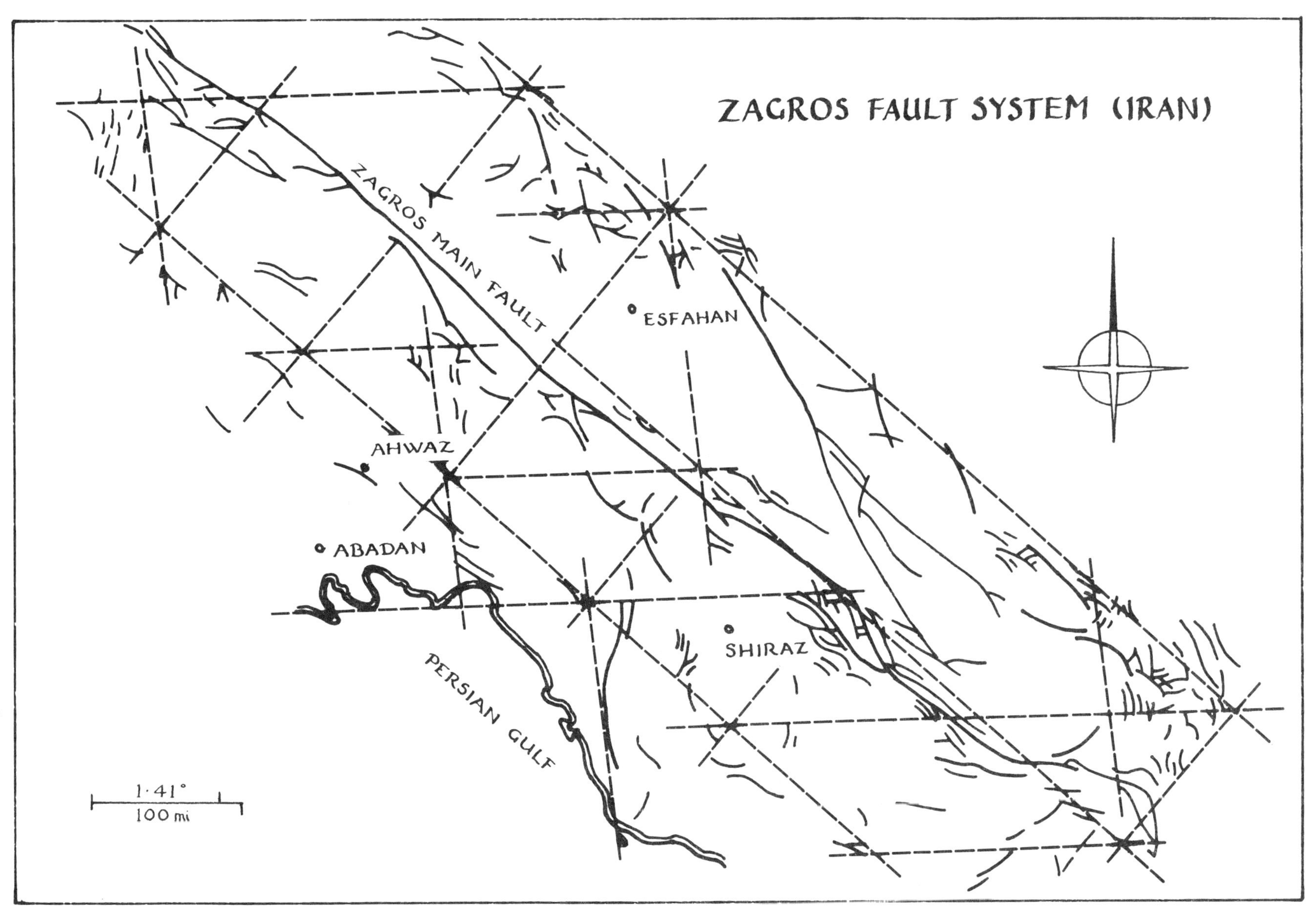

ZAGROS FAULT SYSTEM (IRAN)
ZAGROS MAIN FAULT
ESFAHAN
AHWAZ
ABADAN
SHIRAZ
PERSIAN GULF
1·41°
100 mi

7-2 South Australia
Joint Controlled Pans
What Are The Limits?

This shows a group of joint controlled pans found on the Eyre Peninsula in South Australia. The pans are erosional depressions scoured out of the rock, and although the scouring mechanism is not understood, the pattern is controlled by the joint system in the granite. Joints are small cracks in rocks that commonly form a very regular pattern of lines that interconnect in a rectangular or triangular network. In this figure the erosional pans outline a joint system that forms a rhombus network corresponding closely with the directional pattern. The interval between directionals is 0.309 sec, which measures about 25 feet and is the 23nd member of the hierarchical series given in chapter 2. This represents the smallest interval that could be found in available literature where orientation was given. The largest interval shown in the book is the Pacific Trench System from chapter 3. This system covered the entire Pacific Ocean, with an interval of 11.2 deg which measured about 750 miles at the center of the figure. Based on degrees, the 11.2 deg Pacific Trench interval is more than 100,000 times the 0.309 sec granite pan interval, representing five orders of magnitude.

The patterns found in nature are valid only within certain orders of magnitude, and there are good reasons for this. The defined elements of nature rarely increase in the same proportion, so that a small undetectable error in an element may blossom into a large unacceptable error as order of magnitude increases. A good example of this is the range of validity for Newtonian mechanics, which works well in the practical every day world, but breaks down as velocities approach the speed of light. Nature also often performs differently, depending on the number of items or amount of matter involved. A major problem in engineering concerns the validity of laboratory tests from small samples when applied to the huge masses of material in nature. In oil reservoir analysis, the flow characteristics of the reservoir rock are measured in the laboratory from samples that can be held in a hand. However, the actual reservoir of rock could be several hundred feet thick and extend for miles. When applying the small sample properties to the huge reservoir, it is almost always necessary to make major adjustments in order to match the measured flow history of the oil wells. A third order of magnitude problem is the tendency of natural processes to suddenly shift or change when some threshold is reached. When certain combinations of temperature and pressure occur, substances can shift between vapor, liquid, and solid. If an increasing force is applied to rock, it may for a time deform plastically like a rubber ball, but then suddenly break apart when the applied force reaches a certain limit. The relationship between an applied force and the tendency to deform is also dependent on time. A quick hard blow tends to fracture while a slow steady pressure tends to deform. Ice will shatter when hit with a hammer, but it will flow slowly downhill as a glacier under the steady force of gravity.

This means that manmade discrete mathematical patterns can be fitted over only a part of a continuous and very complex nature. A pattern that works perfectly and has a perfectly logical explanation when applied to a particular set of observations, may fail completely when tested against other sets. This does not necessarily destroy the usefulness of the pattern if the limits or boundaries to its validity can be found. The various equations in the technical and engineering world all have limited application, and it is not unusual to use safety factors of two or more to make sure the relationship is valid. Consideration must be given not only to the size, composition, and shape of the material used, but also to temperature, pressure, and the surrounding environment. For moving parts, the velocity, acceleration, and harmonic vibration must also be considered. Each element in an equation has certain limits of validity, and each must be checked in any engineering analysis. The complexities of nature do not allow for the simple application of an equation or computer program where numbers are fed in on one side and a so-called answer churned out on the other. Without a basic understanding of the limits of usefulness, an analysis method may do more harm than good by giving an appearance of reason, logic, and sophistication that may not exist. This aspect of analysis may be the one most overlooked in today's world of lightening fast computer calculations.

As applied to the Earth as a whole, the comments above take on added meaning. The Earth is a highly complex body in composition, structure, and motion with much that is unknown, and is one of numerous unique bodies that make up the Solar System. The physical variables of the Earth and its companions cover large orders of magnitude, and any single pattern such as presented here must have order of magnitude boundaries that restrict its validity. However, no examples have yet been found to indicate what these might be, even though about five orders of magnitude have been covered. This question of magnitude limits, and other possible constraints and restrictions, is considered one of the most important, but answers will require a far more detailed examination than presented here.

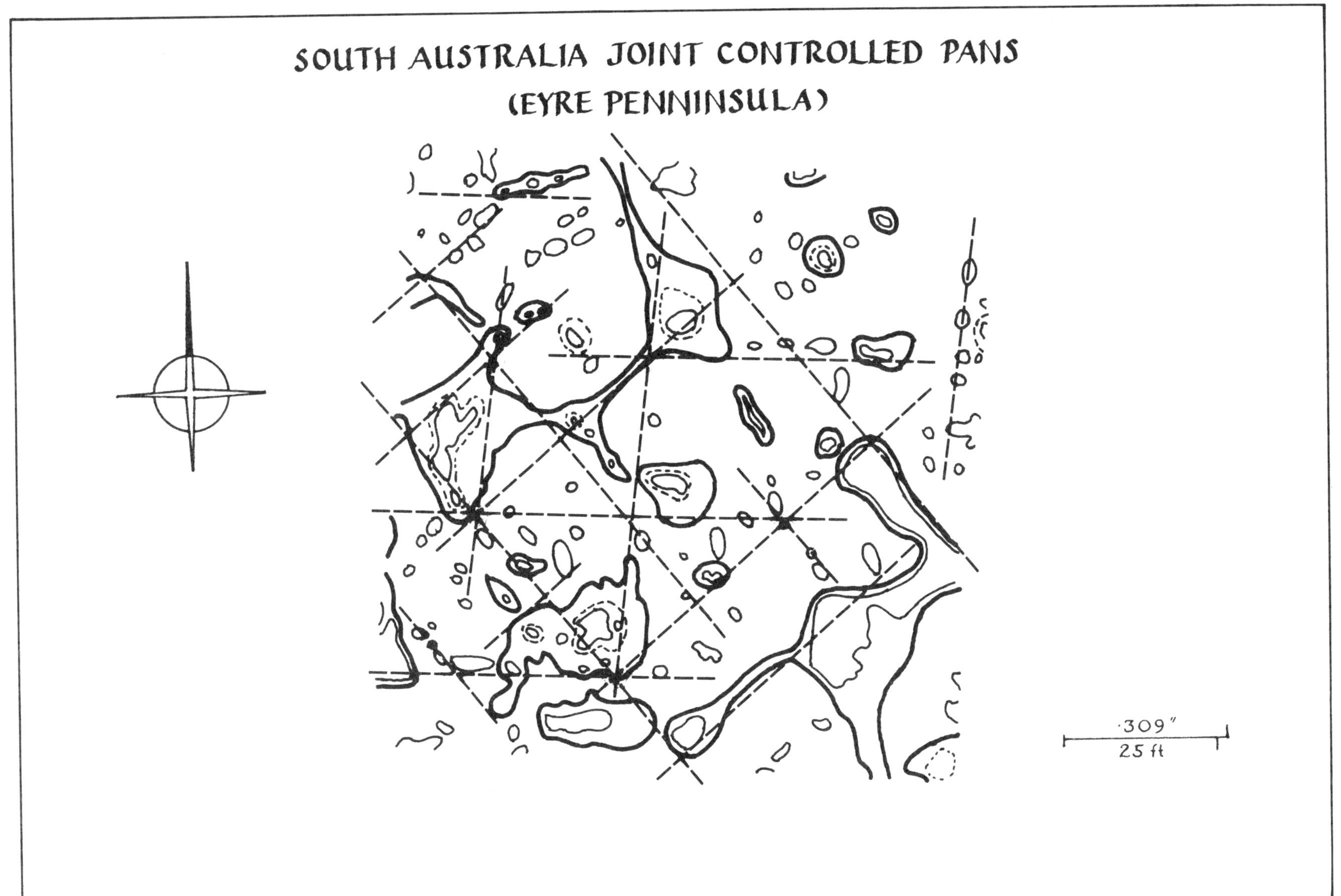

SOUTH AUSTRALIA JOINT CONTROLLED PANS
(EYRE PENNINSULA)
·309"
25 ft

7-3 Yugoslavia Plitvice Lakes
Will other landforms fit?

This shows two of the 16 lakes in Yugoslavia's Plitvice Lakes National Park, located about 50 miles inland from the Adriatic Sea that separates Italy from Yugoslavia. The lakes stretch out along the Matica River from its source in the mountains down to the Kozjak and Oradinoval Lakes that are the largest in the chain. The directional pattern shows a very good correspondence with the lake shorelines which trend N40W along Lake Kozjak in the north. Just below Lake Kozjak the trend shifts to N50E along the upper third of Lake Oradinoval, and then to N7E along the lower two-thirds. Near the middle of this lower portion is an offset across a 19.8 sec hierarchical interval, and at the extreme south is a foot that strikes off to the west along W2N. These lakes were formed somewhat differently from the landforms previously discussed.

Geologic landforms are usually classified based on the process of formation, and karst topography is descriptive of a particular region in Yugoslavia called Kras that lies along the Dalmatian Coast of the Adriatic. In this area there are many small depressions or sinkholes that often are filled with water forming ponds or bogs. Streams in the area tend to be discontinuous and may disappear into a depression, often to reappear lower down as a spring. Much of the area is formed from layers of limestone rock that tend to dissolve slowly in water that is slightly acidic. The limestone is broken up by an intricate network of joints that cut deep into the rock, allowing water to seep down through the cracks and slowly dissolve away the limestone. As the rock dissolves underground, the land above begins to sag down in places, and may eventually collapse to form groups of sinkholes that can either collect and hold water or funnel it down into underground streams. Often caves are formed where the water dissolves out holes in the limestone underground. This type of landscape with depressions, sinkholes, bogs, discontinuous streams, and caves is called "karst topography." The two Plitvice Lakes shown in the figure are large karst landform depressions that have filled with water to form lakes. The directional orientations of the shorelines are probably controlled by the pattern of the joint system that directs the flow of the dissolving water. In the previous figure the directional pattern of the erosional granite pans was also controlled by the joint system, and the strain analogy could explain the similarity of these two joint system patterns on opposite sides of the globe.

Natural landforms rarely agree completely with the simple geologic definitions based on process of formation. The described processes are manmade divisions of a complex interchange of many different forces acting in concert over the entire Earth, and this is complicated by the often changing geologic concepts that describe the formation processes of the various landforms. Consequently, the divisions between landforms are always somewhat cloudy and often confusing. However, this need not distract from the usefulness of the classifications as long as the hazy relationship to the natural world is understood. A strain pattern within the Earth's crust would probably be reflected in some way within any landform formed by any combination of processes. However, considerable variation in the correspondence might be expected, depending on the materials and processes involved. The reaction of rock to a force depends on a complex of factors including composition, time, amount of material, temperature, pressure, and other forces that might be involved. Several examples of different landforms have been given, and each reflects the directional pattern in its own unique way. Continental coastlines tend to follow the directionals, but often show abrupt offsets across hierarchical intervals. Various landforms on the seafloor also seem to follow the directionals, as shown by the India Ocean ridges, the stair step Red Sea Rift, and the Hawaiian chain along with nearby linear seamount chains. Fault systems fit within the directional framework somewhat differently. Although a few faults tend to follow directionals, the smaller faults rarely show a directional trend. Rather, fault ends, breaks, and direction changes were found to be positioned on directionals, as might be expected from mechanical considerations. Flow systems of both streams and volcanics fit within the directional framework with a mix of the above characteristics. The flow channels are often constrained along directionals, but with headwaters, branches, and abrupt changes often located on directionals. The joint controlled patterns of granite pans and karst lakes also tend to follow directionals.

There are a wide variety of landforms, including desert topography, glacier features, and mountain terrain. Although examples of these have not been shown, they would probably also reflect the directional pattern because they have been subjected to many of the same forces. The specific characteristics of the correspondence can be determined only by much additional study related to the specific landform involved.

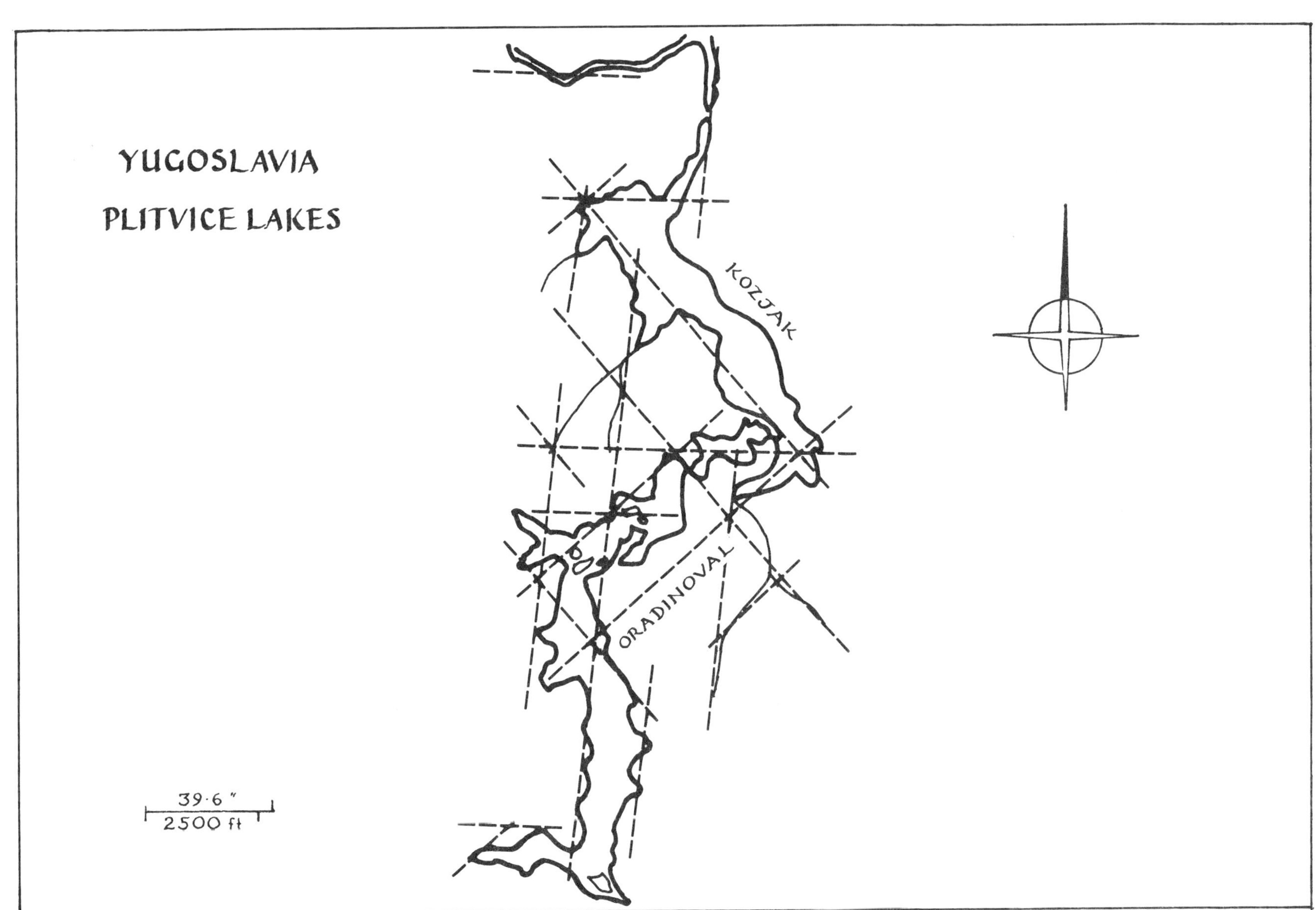

YUGOSLAVIA
PLITVICE LAKES
KOZJAK
ORADINOVAL
39·6"
2500 ft

7-4 World Mercator Map
What About Plate Tectonics?

This shows an enlarged version of the Mercator map given in chapter 2. Coastlines of continents and islands are represented by solid lines. The deep ocean trenches discussed in Chapter 3 are shown as sawtooth lines, and the ocean ridges and rises described in chapter 4 as dotted lines connecting groups of short transverse fault lines. The dotted ridge lines are extended into the Red Sea and East Africa to mark major rift zones, and another probable rift zone across central Asia is shown as a curved dashed line. North of Iceland the probable trace of the Jan Mayan Ridge is also shown by a dashed line. A directional pattern with a hierarchical interval of 11.2 deg has been fitted over the map, showing many directionals coincident with both individual features and with chains of various features. One clearly defined N40W directional follows a feature chain just west of the American Continents. On the north it traces the California coast, proceeds southwest along the South American Peru Coast, and then on to the South Sandwich Trench. Another long N50E directional trend begins along the Taymyr Peninsula in north Russia, then runs parallel to the European coast, continues southwest along the northwest African coast, then crosses South America from the Amazon mouth to the bend in the Peru-Chile Trench, and ends along a portion of the East Pacific Ridge. East of Asia another long N50E directional can be traced down the Kuril Trench and the southwest coast of Japan, and then extended to follow the Southwest Indian Ocean Ridge below Africa.

Many other somewhat shorter trends can be seen. A N50E directional that follows the East African Rift can be extended northeast to trace a probable crustal rift across central Asia. The N7E directional lying on the Ninety East Ridge in the Indian Ocean can be extended north to the intersection of the two diagonal directionals that outline the tip of the Taymyr Peninsula. The two W2N directionals bordering the

Great Lakes of North America can be extended east to also border the Black and Caspian Seas in Europe. Overall, the directional pattern provides a rather tightly fitting network for the features displayed on the map. It seems somewhat surprising that the directional pattern, based on a static Earth observation, should correspond closely with the changing major features that form an integral part of the Plate Tectonic Theory.

Although many of the details remain unresolved, the overall picture of Plate Tectonics is widely accepted by the scientific community. In this theory the upper crust of the Earth is divided into several large semi-rigid plates that are bounded by ocean ridges and trenches. These boundaries are rather spongy and soft in comparison to the ridged plate interiors. Along the ocean ridges molten material from the Earth's interior rises up to the surface and oozes out to continually add new ocean floor area to the plates on either side of the ridge. Along the ocean trenches ocean floor area is lost as the crustal plate on one side of the trench slides down under the plate on the other side, to eventually melt and merge with the molten material in the Earth's interior. These two processes compensate each other so that the total area of the Earth remains unchanged, with the surface gained at the ridges equivalent to the surface lost at the trenches. However, this is not true of an individual plate that changes in area and shape, depending on the different rates that area is gained and lost at the ridge and trench boundaries. The change in plate area is most evident along the trenches where area is being lost on one side as the lower plate slides down into Earth, while area is neither gained or lost on the other side as the upper plate holds a static position along the trench. The net result over the Earth's surface is a complicated interactive network of plate cells that individually change shape and area continually, while the surface of each cell shifts like a treadmill from ridge to trench, carrying with it any continental mass that happens to be stamped on its surface. It is these shifting ridge, trench, and conti-

nent boundaries that seem to fit the static directional pattern.

A meaningful discussion of a changing pattern requires a fixed frame of reference, and for the Earth, the axis of rotation seems the only practical orientation reference. Other reference frames have been suggested based on the earth's magnetic field or core circulation, but none have yet proven acceptable due to the many uncertainties and unknowns involved. Although the axis may wobble a bit, on the average it represents a reasonably steady reference that has been studied in detail over many years. With the rotational axis assumed as reference, the north and south poles are immovable by definition, and the concept of shifting rotational poles is meaningless. It is the plate boundaries and continental coastlines that shift relative to the static axis. The directional pattern and strain analogy both assume the static axis as a reference, and the close fit with plate boundaries could perhaps indicate that these also might best use the same frame of reference.

In the first discussion of this chapter it was suggested that the directional pattern might reflect the end imprints of a torsional oscillation, twisting alternately clockwise and counterclockwise. As the strain pattern changes in tune with the twist, it may be that the plate boundaries are dragged along forcing the area and shape to shift. A simple analogy would be a loosely woven net stretched around a basketball and twisted back and forth. The changing shape of the spaces in the net would represent the changing plate cells on the Earth. This is of course speculation, but it does indicate that the directional pattern and Plate Tectonics are not necessarily in conflict.

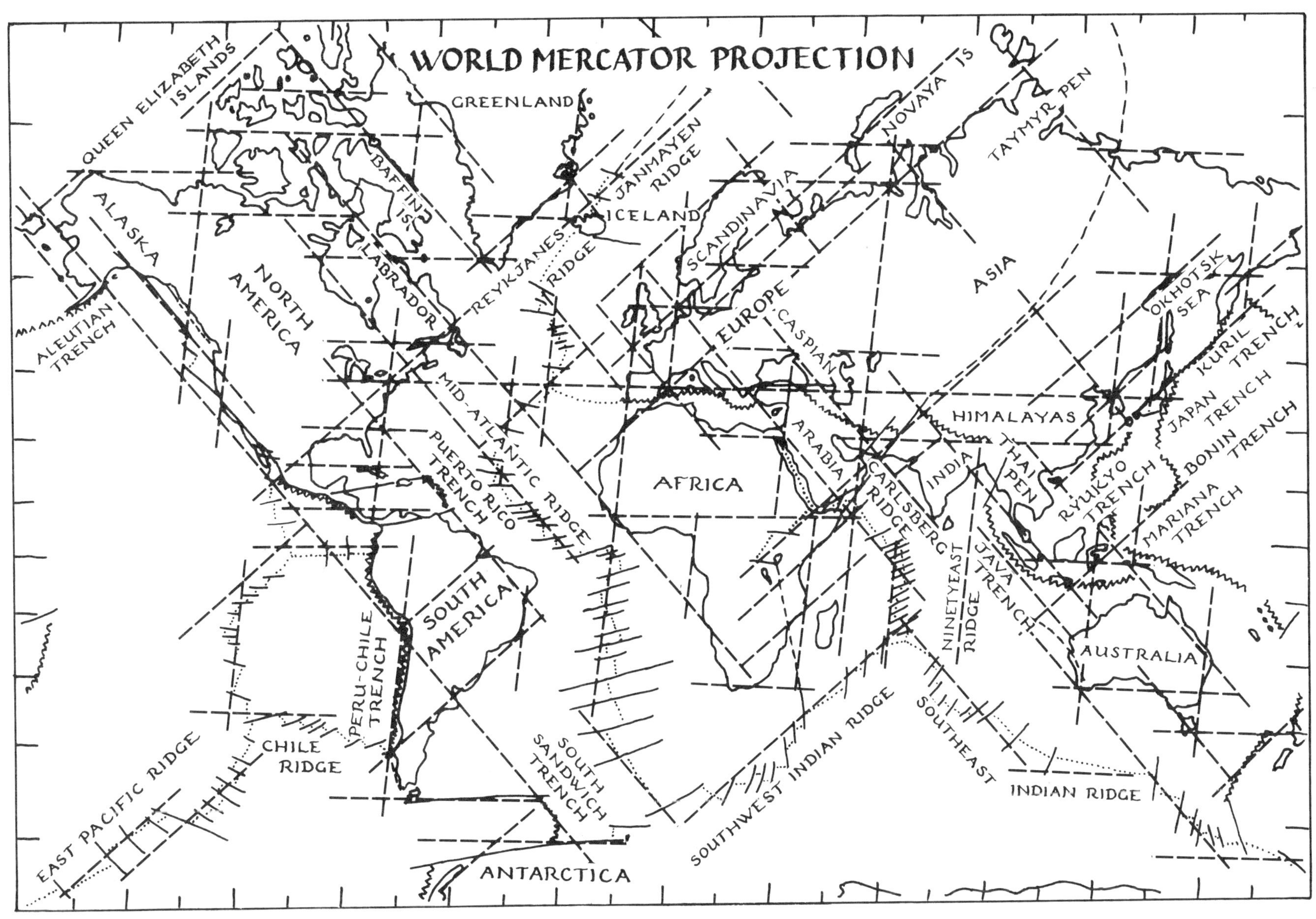

WORLD MERCATOR PROJECTION
QUEEN ELIZABETH ISLANDS
GREENLAND
ALASKA
BAFFIN IS
LABRADOR
NORTH AMERICA
ALEUTIAN TRENCH
REYKJANES RIDGE
JANMAYEN RIDGE
ICELAND
SCANDINAVIA
EUROPE
NOVAYA IS
TAYMYR PEN
ASIA
CASPIAN
OKHOTSK SEA
KURIL TRENCH
JAPAN TRENCH
HIMALAYAS
MID-ATLANTIC RIDGE
PUERTO RICO TRENCH
AFRICA
ARABIA
CARLSBERG RIDGE
INDIA
THAI PEN
RYUKYU TRENCH
BONIN TRENCH
MARIANA TRENCH
SOUTH AMERICA
PERU-CHILE TRENCH
NINETYEAST RIDGE
JAVA TRENCH
AUSTRALIA
CHILE RIDGE
SOUTH SANDWICH TRENCH
SOUTHWEST INDIAN RIDGE
SOUTHEAST INDIAN RIDGE
EAST PACIFIC RIDGE
ANTARCTICA

7-5 *Pacific Ocean Magnetic Anomalies Why 7.2 degrees?*

This shows the striped pattern of reversed magnetic polarities on the Pacific Ocean floor just off the west coast of North America. The Earth is a huge magnet with its north and south magnetic poles located close to the rotational poles. The north pole of a magnetic compass needle points north because like poles attract, and it is pulled toward the north magnetic pole of the Earth. The Earth magnet also acts on lava when it cools down and hardens. As small grains of magnetic iron minerals in the lava begin to solidify, they tend to line up with the Earth magnet. This magnetizes the lava rock with its north polarity pointing toward the Earth's magnetic north pole.

In the early 1960's the magnetic polarities along an ocean ridge were first mapped, and this figure shows the surprising pattern that was found. The stripes represent alternate bands with completely reversed polarity on the sea floor running parallel to the ocean ridge. In one set of alternate bands the north polarity of the rock points towards the Earth's north magnetic pole, and in the other alternate set the north polarity points towards the Earth's south magnetic pole. As described on the preceding page, lava from the Earth's interior is oozing out along the ocean ridges to solidify and form new ocean floor, and as explained above, the lava is magnetized as it hardens with its polarity in line with that of the Earth. The only reasonable explanation for the alternate bands of reversed polarity along the ocean ridge is that the Earth's north and south magnetic poles have switched places from time to time in the past. This means that sometimes the Earth's north magnetic pole is near the north rotational pole as it is today, but at other times it lies near the south rotational pole. The discovery and explanation of periodic reversals in the Earth's magnetic poles was critical to the acceptance of Plate Tectonics by the geologic community.

A directional network with a 2.81 deg hierarchical interval has been fitted over the magnetic polarity pattern. Although the individual stripes generally follow N7E, the overall pattern resembles a wide band that has been bent at the center to form a knee. The band parallels the west coast of North America which abruptly shifts direction at the southern tip of Vancouver Island, trending N40W to the north and N7E to the south. North of the kneebend the magnetic pattern is enclosed in a rectangle with the long sides running N40W and the short sides N50E. The N50E directional that marks the southeast side of the rectangle follows a line of offsets in the stripes, and can be extended to the intersection of the two directionals that follow the North American coastline. The correspondence of the directional pattern and the magnetic pattern should not be all that surprising. In the previous figure it was shown that the ocean ridges generally follow directionals, and the magnetic stripes are aligned with the ocean ridges. Of course, this is purely observation with nothing to explain why this particular set of directional angles seem to fit so many of Earth's features.

The directional pattern was explained with a strain analogy in which the Earth was twisted around its rotational axis to create a rhombus strain pattern in the crust. On Mercator Projection the extreme strain lines are represented by the straight sides and diagonals of the strain rhombus. However, the twisting force required to create the strain pattern would have some very unique properties. In applying the force there are two major considerations. First, there must be some positive connection between the force and the object it is acting on, and second, the force must be generated in some way. An example illustrating this would be the twisting of a flexible willow branch. The hands must grasp the branch firmly at either end so that friction between the hands and the branch connect the force to the object. The hands are then rotated in opposite directions, and the amount of twisting depends on the amount of force applied up to the point where the twist is enough to

split and tear the branch apart. As applied to the strain analogy, some presently unknown force must be connected in some way to the Earth's rotational poles, and must act to rotate the Earth in opposite directions. In addition, the magnitude of the force must be just right to generate and hold a 7.2 deg twist. However, the problem is even more complicated because there are changes with time. If a steady force acts over a long period of time, a material will tend to rearrange and adjust itself to compensate for the force. This internal adjustment allows for twisting to continue without stopping at some particular angle. All these unanswered questions indicate that the strain analogy must have some rather severe limitations.

No analogy can be applied indiscriminately to the complexities of nature. At best, it not only provides a limited insight into a natural process, but also points out inconsistencies and anomalies that define the boundaries of meaning. Continuing study may resolve certain difficulties and extend the boundaries of the analogy, or possibly bring out new ideas that provide a more complete explanation of the observations. In the later case the original analogy would be discarded and replaced with the new explanation. The strain analogy, with all its deficiencies, will hopefully serve as a starting point for further study from which a better understanding of the directional pattern observation might emerge.

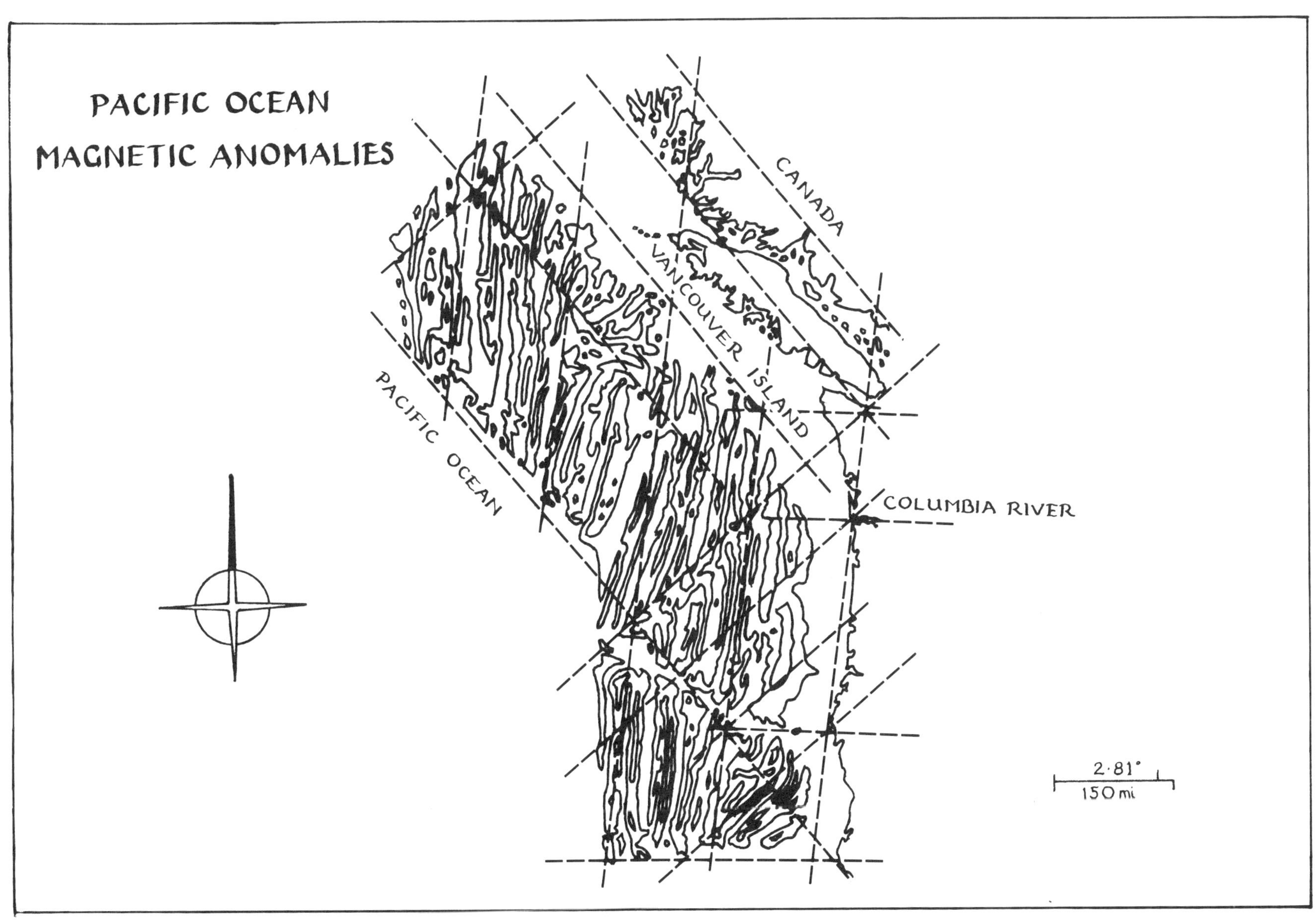
PACIFIC OCEAN
MAGNETIC ANOMALIES
CANADA
VANCOUVER ISLAND
PACIFIC OCEAN
COLUMBIA RIVER
2.81°
150 mi

7-6 England Stonehenge Where Do We Go From Here?

This shows a plan of Stonehenge, a New Stone Age temple located in southern England. The overall pattern is formed by a group of concentric circles cut by an avenue striking off towards the northeast. The three concentric circles are marked by filled in pits that are represented by small ovals drawn with the lighter lines. The outermost circle is ringed with 56 pits called the "Aubrey holes," then comes the next smaller circle of 30 "Y" holes, and inside that is another of 30 "Z" holes. Inside these three outer circles is another group of smaller circles originally formed of large upright stones that are represented by ovals drawn with heavier lines. Many of the larger ovals show stones that have fallen over. The outer stone circle was made of large upright sandstone blocks topped by a raised stone pathway that ran around the circle. It is called the "Sarsen Circle" from the type of sandstone used. Just inside the Sarsen Circle is another of smaller stones called the "Bluestone Circle," and at the center is a horseshoe-like partial circle opening out towards the northeast avenue. This inner horseshoe was originally made of five pairs of large Sarsen stones with a stone lintel resting on top of each pair. And finally, just inside the Sarsen horseshoe is another matching horseshoe of bluestones. The other heavily outlined stones are called the station stones and are found on the outermost pit circle, with several more located along the middle of the northeast trending avenue. This Stone Age temple of pits and large stone blocks was constructed in phases over a period of more than 1,000 years, beginning about 4,800 years ago.

A directional pattern with a 0.618 sec hierarchical interval shows a surprisingly good match when fitted over the Stonehenge plan. The avenue is bordered by two N50E directionals separated by a 1.24 sec interval. These directionals can be extended down to the left where they trace the sides of the Sarsen horseshoe. The two station stones on the lower right of the outermost Aubrey pit circle mark the short side of a rectangle inscribed within the Aubrey circle. The shorter sides of the rectangle trend N50E and span a 3.47 sec interval, while the longer sides run N40W with a length five times the shorter sides. These rectangle sides not only represent the diagonals of the strain rhombus, but also approximate certain astronomical sight lines. The short N50E side, in common with the avenue, points towards the midsummer sunrise in the northeast, and the midsummer sunset in the opposite direction, while the long N40W side points towards the moonrise at major standstill in the southeast, and moonset at major standstill in the opposite direction. A major standstill marks the farthest the moon shifts to the north or south on the horizon during its 18.6 year cycle. There are undoubtedly those who would attribute the close correspondence of astronomical sight lines, the directional pattern, and an ancient manmade structure to little green spacemen of vast knowledge who came down to Earth some 5,000 years ago to instruct our forefathers in astronomy, geology, and architecture. However, there is an alternate explanation that perhaps makes a good deal more sense, given a reasonable respect for our ancestors' brains.

In the strain analogy, the directional pattern represents the extreme strain lines that develop from a twisted earth, and these strain line directionals mark preferential flow paths. Any liquid such as water or lava would tend to follow these paths, creating erosional and volcanic topography reflecting the directional pattern. Faults and fractures would often start and end along these strain lines, further complementing the pattern. All Earth processes would be constrained to a degree by the strain network, so that the overall grain or lay of the land would tend to reflect the directional pattern. This land grain is a major constraint in the location and orientation of man's construction, so that houses and other structures would tend to follow the lay of the land, again reflecting the directionals. Another item of importance for the ancients of southern England is that they just happened to live at a latitude where the diagonal N40W and N50E directionals closely matched the orientations of the midsummer Sun and the major standstill Moon.

Once upon a time many thousands of years ago, there was a boy full of curiosity, imagination, and wonder who lived in a hut in southern England. He didn't fully respect the wisdom of the elders, and sometimes would actually argue with them, disputing their true and accepted explanations. This made others uncomfortable, so they stayed away and left him to ponder alone on the beautiful complexity of nature. One day during the midsummer celebrations, when everybody else was feasting, dancing, and laughing, he sat alone, resting against the corner of his hut taking in the sunset. He watched and wondered as the sun dipped down toward the horizon and the colors in the sky shifted and changed. But then, suddenly, he saw something that he had never noticed before. The setting midsummer Sun was lined up along one side of his hut. This was fascinating, and he wondered if the same thing happened with all the other huts, so he jumped up and quickly ran round the village to find most other huts also lined up with the setting Sun. In other years he watched the midsummer Sun from other villages and saw much the same thing. He told no one of his discovery, except his children who passed the secret on to their children, and eventually it became the forbidden knowledge of the elders who decreed that all temples be oriented towards the midsummer Sun. Many generations later that decree established the orientation of Stonehenge, the largest temple of all — or something like that.

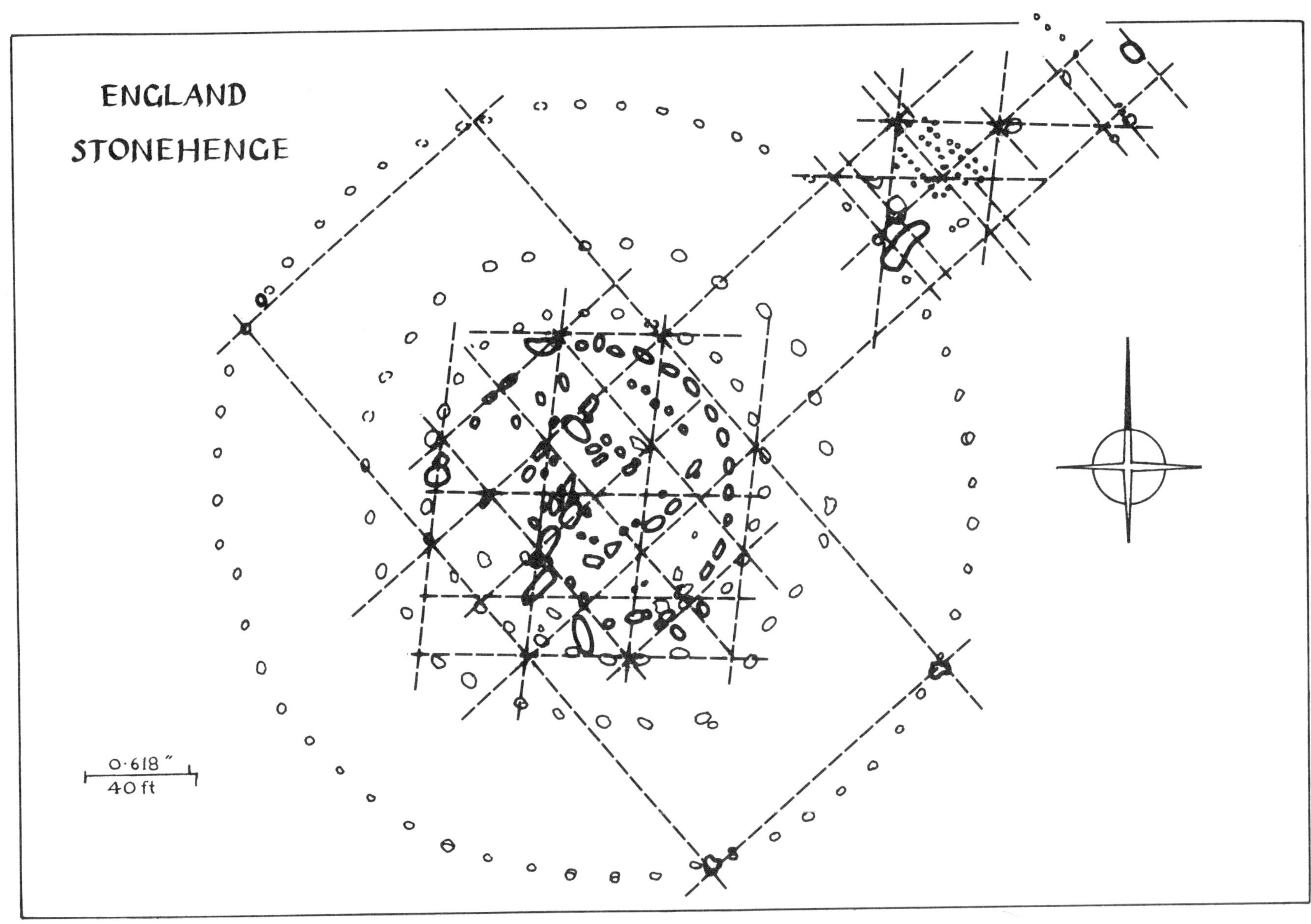

ENGLAND
STONEHENGE
0·618"
40 ft

Geologic analysis is many things, but explicit, routine, and straightforward it certainly is not. No one overall theory has ever proven universal in its application. The new Plate Tectonics brings much together, but it also has its contradictions and questions. The driving force of the shifting plates has yet to be fully explained, and the paths taken by the plates over geologic time is still conditioned with conjecture. In addition, there is no evidence of plate movement on other Solar bodies, so that their surface patterns must be explained by other processes. Nature does not act in discrete ways on defined groups of items. The distinct boundaries that divide nature into separate categories are manmade divisions created for man's analysis. Nature operates in a many-faceted orchestration of overlapping processes over an intricate field of interconnected objects. The complex harmonics merge and blend as dominance of process and object shifts and changes. This is the wonder of nature that demands the consideration and respect of every analyst. Final answers and complete explanations are not to be expected from man's analysis of nature's eccentricities. With the continued analysis of new observations, the old questions may become better understood, but a multitude of new questions continue to emerge. Analyzing nature is wonderful and fascinating, but an endless quest. There will always be questions on which to ponder and work.

However, perhaps the greatest wonder is that orderly patterns do appear in the complex kaleidoscope of nature. The daily and seasonal cycles are obvious to everyone. The day has its Sun and the night its stars. The Moon cycles through its monthly phases, and the Sun marks out a year for each circuit through the stars, while both swing back and forth across the horizon. The Sun repeats its horizon shift each year and the Moon each month, but with the full Moon at one extreme in summer and at the other extreme in winter. The planets move through the stars along the Sun's path, each with its own unique cycle. Venus is quite regular, completing five cycles in three years, while Jupiter and Saturn also keep a steady beat of 12 and 30 years. The sea has its tidal cycles and weather generally follows the seasons. Then there is life itself that is an integral part of nature. Sea life is often in tune with the tides, and birds migrate seasonally. Many insects and plants come to life in the spring after lying dormant throughout the winter. Our ancestors observed with wonder this cyclic dance of elements that involved all of nature, both animate and inanimate, and they pondered on the cycle relationships. They mentally counted, collected, and sorted, then meshed it all together in elaborate, intricate, and interconnected schemes of thought that were the models of yesterday. The mental development of unobserved relationships between observed patterns requires imagination, that elusive thought process of man that creates the explanations of the unexplainable world.

For most people explanation is a satisfying necessity that defines the regularity of confusing nature. It is a product of imagination, constrained by accepted cultural values. Man has always searched for answers to the disparities and inconsistencies of the unfair natural world. Why is the redbird red? Because he was burned red when he brought fire down to man, or perhaps because he can better attract the female during the spring mating season. Both reasons are imaginative and satisfying, but both are also consistent with the cultural tide of that particular age. Our ancestors were quite satisfied with the cycle on cycle explanation of the natural world, and it did include nearly all that was known at the time. Today, explanation is often provided by sophisticated mathematical computer models that also include nearly all that is known at this time. Both old and new explanations are imaginative and satisfying to those involved, but just as the ancient cyclic system was replaced, so also will the new models require adjustment and change with the acquisition of new information. On a smaller scale, each analysis starts with a set of known observations, and imagination must build the interdependent network to contain those observations. Today's analyst has many formulas and models to work with, but rarely does any one provide a complete solution. Each set of observations demands an approach that is unique in some way, and it is imagination that brings together the final working pattern.

However, there is an important difference between our ancestors' accepted schemes of interwoven cycles and the many new explanations that pour out of analysis today. A new idea must now be tested against established patterns and repeated observations before it gains respectability. This requires development in sufficient detail to provide a pattern that can be tested, which in essence is the difference between speculation and theory. Many brilliant and beautiful ideas have appeared in pseudo scientific literature that must remain speculative because there is no way to test compatibility with the universe of observation. They are not to be discarded, because they represent the wonder of nature integrated with man's imagination, and with further development a few may eventually earn the designation "theory." However, most will remain in the fascinating world of science fiction.

The pattern and explanation described in this book has yet to be developed to the degree that would define it as a theory, and for the present it must remain merely observation and analogy. At best, statistical testing might provide a degree of confirmation, but eventually it must be developed in more detail and adjusted to remove the many inconsistencies. This will undoubtedly require considerably more study by more knowledgeable analysts. For now, it is hoped that the pattern will serve as a useful framework on which to build.

BIBLIOGRAPHY

Ager, Derek V., *The Nature of the Stratigraphical Record,* Macmillan Press, Ltd., London, 1973.

Armstrong, R. Warwick, *Atlas of Hawaii,* The University of Hawaii Press, Honolulu, 1983.

Baker, Victor R., *The Channels of Mars,* University of Texas Press, Austin, 1982.

Beatty, J. Kelly, O'Leary, Brian, and Chaikin, Andrew, *The New Solar System,* Cambridge University Press, Cambridge, 1982

Beckwith, Martha, *Hawaiian Mythology,* University of Hawaii Press, Honolulu, 1970.

Bott, Martin H., *The Interior of the Earth: its Structure, Constitution and Evolution,* Elsevier Science Publishing Co., New York, 1982.

Bradford, Ernle, *Mediterranean, Portrait of a Sea,* Hodder and Stoughton, London, 1971.

Carey, S. Warren, *The Expanding Earth,* Elsevier Scientific Publishing Co., Amsterdam, 1976.

Carr, Michael H., *The Surface of Mars,* Yale University Press, New Haven, 1981.

Carr, M.H., and Greeley, R., *Volcanic Features of Hawaii, A Basis for Comparison with Mars,* NASA, Washington D.C., 1980

Decker, Robert, and Decker, Barbara, *Volcanoes,* W.H. Freeman and Co., San Francisco, 1981.

Derry, Duncan R., *World Atlas of Geology and Mineral Deposits,* Mining Journal Books, Ltd., London, 1980.

Frazier, Kendrick, *Our Turbulent Sun,* Prentice-Hall, Inc., Englewood Cliffs, 1982.

Graves, Robert, *The Greek Myths,* George Braziller, Inc., New York, 1959.

Halbouty, Michel, ed., *Geology of Giant Petroleum Fields,* American Association of Petroleum Geologists, Tulsa, 1970.

Hazlett, Richard W., *Geologic Field Guide, Kilauea Volcano,* Hawaii Volcano Observatory, Hilo, 1987.

Heggie, Douglas C., *Megalithic Science,* Thames and Hudson, 1981.

Iacopi, Robert, *Earthquake Country,* Lane Publishing Co., Menlo Park, 1980.

Klepesta, Josef, and Rukl, Antonin, *Constellations,* Hamlyn House, London, 1969.

Le Pichon, X., Francheteau, J., and Bonnin, J., *Plate Tectonics,* Elsevier Scientific Publishing Company, Amsterdam, 1973.

Lipman, Peter W., Mullineaux, Donald R., *The 1980 Eruptions of Mount St. Helens, Washington,* United States Geological Survey, Washington D.C., 1981.

McClelland, Peter D., *Causal Explanation and Model Building in History, Economics, and the New Economic History,* Cornell University Press, Ithaca, 1975.

Macdonald, Gordon A., Abbott, Agatin T., and Peterson, Frank L., *Volcanoes in the Sea, The Geology of Hawaii,* University of Hawaii Press, Honolulu, 1983.

McWhirter, Norris, *Guiness Illustrated Encyclopedia of Facts,* Bantam Books, New York, 1981.

Marsh, James H., ed., *The Canadian Encyclopedia,* Hurtig Publishers Ltd., 1985.

Morison, Samuel Eliot, *The Oxford History of the American People,* Oxford University Press, New York, 1965.

Morrison, David, and Samz, Jane, *Voyage to Jupiter,* NASA, Washington D.C., 1980.

Morrison, David, *Voyages to Saturn,* NASA, Washington D.C., 1982.

Murray, Bruce, Malin, Michael C., and Greeley, Ronald, *Earthlike Planets, Surfaces of Mercury, Venus, Earth, Moon, Mars,* W.H. Freeman and Co., San Francisco, 1981.

Myles, Douglas, *The Great Waves,* McGraw-Hill Book Co., New York, 1985.

Noyes, Robert W., *The Sun, Our Star,* Harvard University Press, Cambridge, 1982.

Rukl, Antonin, *Moon Mars and Venus,* Hamlyn Publishing Group Ltd., London, 1976.

Rybach L., and Muffler, L.J.P., eds., *Geothermal Systems: Principles and Case Histories,* John Wiley and Sons, Ltd., Chichester, 1981.

Santillana, Giorgio de, and Dechend, von Hertha, *Hamlet's Mill,* Gambit, Ipswich, 1969.

Shelton, John S., *Geology Illustrated,* W.H. Freeman and Co., San Francisco, 1966.

Shipley, Joseph T., *The Origins of English Words, A Discursive Dictionary of Indo-European Roots,* John Hopkins University Press, Baltimore, 1984.

Stacey, Frank D., *Physics of the Earth,* John Wiley and Sons, Inc., New York, 1977.

Stearns, Harold T., *Geology of the State of Hawaii,* Pacific Books, Palo Alto, 1985.

Tributsch, Helmut, *When Snakes Awake,* The MIT Press, Cambridge, 1982.

Uyeda, Seiya, *The New View of the Earth,* W.H. Freeman and Co., San Francisco, 1971.

Vitiliano, Dorothy B., *Legends of the Earth,* Indiana University Press, Bloomington, 1973.

Wilford, John Noble, *The Mapmakers,* Alfred A. Knopf, New York, 1981.

Wood, John Edwin, *Sun, Moon and Standing Stones,* Oxford University Press, Oxford, 1978.

Wood, Robert Muir, *The Dark Side of the Earth,* George Allen and Unwin, Ltd., London, 1985.

Wylie, Francis E., *Tides and the Pull of the Moon,* Stephen Greene Press, Battleboro, Vermont, 1979.

INDEX

The Twisted Earth
was typeset in Else Medium
by Coman & Associates, Tulsa, Oklahoma
and printed by Walsworth Publishing, Inc,
Marceline, Missouri.

The cover was produced using a
handmade paper sculpture by Dana De Kalb,
San Francisco, California.